BIO-AGENTS FOR HORTICULTURAL CROPS

(Production, Multiplication and Application)

THE EDITORS

Dr. N. B. V. Chalapathi Rao working as Principal Scientist (Entomology), ICAR-AICRP on Biological Control, Dr.YSRHU-Horticultural Research Station, Ambajipeta, Andhra Pradesh. He has identified various predators and parasitoids against insect pests of coconut and developed biocontrol based IPM module against insect pests of coconut. Established biocontrol laboratory, wherein he trained number of scientists on production and mass multiplication techniques of bioagents. He has published 70 Research papers, 2 book chapters, 12 technical bulletins, 61 popular articles and executed Seven externally funded projects. He is the recipient of 29 awards at national and international level. Based upon the work carried out by him and his team, the ICAR-AICRP, Ambajipeta centre has recognized as Best ICAR-AICRP centre at national level.

Dr. B. Srinivasulu presently working as Principal Scientist & Head, Dr. YSRHU-Horticultural Research Station, Ambajipeta, Andhra Pradesh, with 39 years of experience in Research, Extension and Development activities in Horticulture and allied sectors. He has developed biocontrol based IDM package against major diseases in coconut, which has become a national recommendation and also associated in release of 5 varieties in coconut. He is the recipient of 13 awards at national and international level. He has published 6 Books, 8 Chapters in books, 147 Research articles, 60 Technical bulletins, more than 150 Popular articles and executed Nine externally funded projects.

Dr. T. Janakiram presently working as a Vice chancellor of Dr. Y. S. R. Horticultural University, Andhra Pradesh. He developed and released of 6 China aster, 10 Chrysanthemum, 10 Gladiolus varieties and 4 hybrids of Tuberose and 5 gladiolus varieties and 5 chrysanthemum varieties. He is the recipient of 15 awards at national and international level and fellow of six societies. He has guided 2 Ph.D and 8 M.Sc students at IIHR, Bengaluru and IARI, New Delhi. He has outstandingly contributed to horticulture research and development in his career spanning 32 years with impressive record of 260 publications.

BIO-AGENTS FOR HORTICULTURAL CROPS

(Production, Multiplication and Application)

– Editors –

N. B. V. Chalapathi Rao

B. Srinivasulu

T. Janakiram

2025

Daya Publishing House®

A Division of

Astral International Pvt. Ltd.

New Delhi – 110 002

ISBN : ***978-93-5919-481-3***

Publisher's Note:

Every possible effort has been made to ensure that the information contained in this book is accurate at the time of going to press, and the publisher and author cannot accept responsibility for any errors or omissions, however caused. No responsibility for loss or damage occasioned to any person acting, or refraining from action, as a result of the material in this publication can be accepted by the editor, the publisher or the author. The Publisher is not associated with any product or vendor mentioned in the book. The contents of this work are intended to further general scientific research, understanding and discussion only. Readers should consult with a specialist where appropriate.

Every effort has been made to trace the owners of copyright material used in this book, if any. The author and the publisher will be grateful for any omission brought to their notice for Acknowledgment in the future editions of the book.

Published by : **Daya Publishing House®**
A Division of
Astral International Pvt. Ltd.
– ISO 9001:2015 Certified Company –
4736/23, Ansari Road, Darya Ganj,
New Delhi-110 002
Ph. 011-43549197, 8130496929
E-mail: info@astralint.com
Website: www.astralint.com

Dr.Y.S.R. Horticultural University

www.drysrhu.ap.gov

Dr.T.JANAKIRAM, ARS
Ph.D (IARI) FISOH FISVS FHSI & FISHRD
Vice-Chancellor

Message

India is currently producing about 320.48 million tonnes of horticulture produce and has surpassed the food grain production of 315.7 million tonnes. India is world leader in the production of a variety of fruits *viz.,* Mango, Banana, Guava, Papaya, Sapota, Pomegranate, Lime, Aonla and is the second largest producer of fruits and vegetables and maintained its dominance in the production of Coconut, Cashewnut and Spices.

Food plants of the world are damaged by more than 10,000 species of insects, 30,000 species of weeds, 100,000 diseases (caused by fungi, viruses, bacteria and other microorganisms) and 1000 species of nematodes. In India, Annual production losses due to pests are around 30-35 per cent the crop yield. In the present era the proportion of pests and diseases are increasing concurrently and as a result, harmful chemicals were used more frequently to manage them resulting in the increase of species resistant to pesticides and fungicides. Excessive use of chemical pesticides leads to environmental problems causing a matter of concern and in the wake of increased call for pesticide free food approach more emphasis is being given to the use of eco-friendly approaches for the management of insect pests and diseases. In this context, Biological control has received increased attention because to its low cost, effectiveness, and environmental friendliness and there is an immediate requirement to scale up the bio control agents production substantially.

This book deals with mass multiplication techniques of bioagents, biological control based protection technologies of flower crops, plantation crops *viz.,* Coconut, Cocoa, Arecanut, Oil palm, Cashewnut and nematode management. Information related promotion of bio-control research and strategies to overcome the challenges in application of bio control and institutes involved are also included in this book.

I congratulate the all the authors for contributing chapters and the editors, Dr. N. B. V. Chalapathi Rao and Dr. B. Srinivasulu for such an excellent compilation of information on multiplication, production and application of bio agents and bringing out as book. This publication will be a great use for the crop protection students of horticulture, promoters of biological control and stakeholders of Horticulture thereby improving eco friendly environment for a better living and healthy life.

Vice-Chancellor

dministrative Office, Venkataramannagudem - 534101, West Godavari District, Andhra Pradesh, In
Ph : 08818-284312, Mobile : 7382633666, Email : vc@drysrhu.edu.in

Preface

At the very outset, I feel honoured to present this book Bio agents for Horticultural crops (Production, Multiplication and Application), keeping in view of the present importance of conservation of Bio diversity which is threatened by modern agricultural practices particularly due to excessive and indiscriminate use of pesticides . Natural biological control of insect pests by bio agents is a evolutionary process that has been occurring since the beginning of the evolution . I have tried my best to discuss as many contemporary and relevant issues pertaining to biological control in important Horticultural crops particularly plantation and flower crops, the challenges faced in this sector collectively from the research contribution of numerous workers and also a considerable contribution of my own. Besides the information on Nematodes in Horticultural Crops has also been elaborated. Wherever possible, I have also tried to incorporate the latest research findings importantly based on bio control alone . The book contains ten chapters where in the all the aspects pertaining to Production, Multiplication and Application of bio agents in plantation, flower crops and most importantly discussed research strategies to overcome these challenges .

I express my deep gratitude to Dr. T. Janakiram sir and to Dr. B. Srinivasulu garu for their guidance and encouragement to write this book. Special and sincere thanks are due to Dr V. Anoosha for helping in critically reading the manuscript and proofs, designing the covers, and spending her valuable time in checking the scientific names and spellings, for extending help in the collection of bibliographical material and in several other ways. Thanks are also due to Sri. A.Kireeti and Dr .G. Koteswara Rao for their valuable help in critically reading the manuscript and assistance in preparation of this book.The painstaking and moral support extended by my wife Smt. Mortha Gowri and son N. Sai Ashwin are duly and affectionately acknowledged. Thanks are also due to Sri V. Raja Ramesh , Sri P. Rajesh and Smt M. Annapurna who checked and typed the Manuscript .Thanks are due to all the colleagues and friends who always encouraged and stood by me to bring out this book.

N. B. V. Chalapathi Rao

Contents

Chapter-1

BIOLOGICAL CONTROL OF INSECT PESTS -ITS RELEVANCE AND IMPORTANCE

V. Anoosha[1], S. Adarsha[2] and B. Neeraja[1]
[1]Dr YSRHU-Horticultural Research Station
Dr. YSR Horticultural University
Ambajipeta, Dr. B. R. Ambedkar Konaseema District, A.P.
[2]Dr YSRHU-Krishi Vigyan Kendra
Dr. YSR Horticultural University
Pandirimamidi, Alluri Sitharama Raju District, A.P.

Introduction

Since the beginning of farming, insect pests have been dreadful woes to farmers for many decades leading to huge economic loss, and even total crop destruction in severe cases. Due to indiscriminate use of pesticides to control the insect pests leads to environmental pollution. Additionally, there is more number of increasing species that are pesticide resistant. In recent years, the application of biological control approaches for the management of insect pests has received increased attention because to its low cost, effectiveness, and nature. Therefore, biologically based technologies could be more convectional to solve the urgent needs in pest management. Biological control is "conventional use of living organisms to control the pests (including insects, mites, weeds and plant diseases)". It relies on predation, parasitism or other natural mechanisms along with pivotal role of human management. It is effective against all pests, however the techniques and methodologies employed vary depending on the type of pest. Since the beginning, biological control has enjoyed varying popularity of being ecologically safe and effective means of combating pest problems (De Bach, 1964).There are three basic types of approaches of biological control that includes classical biological control, augmentative biological control and conservation.

Relevance

From the last 50 years farmers adopted traditionally used broad-spectrum pesticides such pyrethroids, carbamates, organochlorines, and organophosphates based pest control approaches. Pesticides provide excellent control of the target pest at the same time they are the source for environment pollution, bioaccumulation through food chains, uneconomic production costs and risks for human health (NRC, 1986). Evolution of pest resistance and loss of abundance in insect natural enemies (*i.e.*, generalist predators and host-specific parasitoids) are main disadvantages of chemical control. Further, loss of these natural biotic population mortality factors can give rise to secondary pest outbreaks as well as resurgence of the pest a grower originally intended to control (Stern *et al.*, 1959). In a book written by Rachel Carson (1972) "**The Silent Spring**" showed a way out to look for alternative pest control tactics rather than sole dependent on pesticides which eroded the natural enemies and also polluted the environment. This publication created awareness among the scientific and farming communities on ill effects of chemical pesticides has led to evolve a sound and viable plant protection alternatives rather than sole dependent on chemical pesticides. In India also the ill effects of insecticides is witnessed Kasargod district, Kerala state to contain the tea mosquito bug in cashew orchard. To overcome the problems associated with pesticides the concept of economic threshold level was proposed by Stern and his associates and it gave impetus for the development of a concept called integrated pest management (IPM) was initiated on major crops to combat the insect pests by use all possible management tactics. Biological control is one of the integral components of the IPM which involves the use of any living organism and it's by product to contain other living organism which is a pest. Biological control comprises of natural enemies (predators and parasitiods), and the disease causing microbes in insect pests *viz.*, bacteria, fungi, viruses and nematodes and their by products to regulate the insect pest population (Johnson *et al.*, 2008).

Importance of Biological Control

World population is expected to grow from around 6 billion to 8.3 billion by 2030 and world cereal equivalent (CE) food demand is projected to 10,094 million tones. In the present era the proportion of pests and diseases increased concurrently and as a result, harmful chemicals were used more frequently to manage them resulting in the increase of species resistant to pesticides and fungicides. The use of eco-friendly pesticides for crop production has received increased attention in recent years with the signing of the General Agreement on Trade and Tariff of the World Trade Organisation due to their least hazardous nature, low levels of disease resistance, and minimal residue issues. Biological control has been a precious approach in pest management approaches around the world for long years. This approach can combat arthropod pests that encompass developed resistance to insecticides. Biological control strategies are used all around the world to manage pests in horticultural, agro forestry, and agricultural crops. Methods of biological control are successful in various situations, times and in various places. Recent losses of conservation land and increased use of genetically modified crops, rising demand for organic produce and Natural farming, biological control will have an important role to play and can be used as an integral part of IPM.

History of Biological Control and Classical Examples of Biological Control

The history of the use of Biological control agents way back to as early as 200 AD. While there isn't much written documentation found, but texts in Talmud from 200 AD have mentioned the use of ant colonies against each other to check their populations. In 300 AD, augmentation of Asian weaver ants (*Oecophylla smaragdina* F.) was done in citrus trees of China to protect them from insects that feed on foliage. Various unrecorded use of Bio control agents has been carried throughout centuries.

Table 1: Milestones in history of biological control

1762	-	'Mynah' bird imported from India to Mauritius to control locust.
1770	-	Bamboo runways between citrus trees for ants to control caterpillars.
1888	-	First well planned and successful biological control attempt made
1888	-	Vedalia beetle (*Rodolia cardinalis*) was imported from Australia to control cottony cushion scale, *Icerya purchasi* in California (USA) citrus gardens.
1898	-	First introduction of natural enemy into India
1898	-	A coccinellid beetle, *Cryptolaemus montrouzieri* was imported into India from Australia and released against coffee green scale, *Coccus viridis*. Even today it is effective against mealybugs in South India.
1920	-	A parasitoid *Aphelinus mali* introduced from England into India to control woolly aphid on Apple, *Eriosoma lanigerum*.
1929-31	-	Rodolia cardinalis imported into India (from USA) to control cottony cushion scale *Icerya purchasi* on Wattle trees.
1958-60	-	Parasitoid *Prospatella perniciosus* imported from China. Parasitoid *Aphytis diaspidis* imported from USA Both parasitoids used to control Apple Sanjose scale *Quadraspidiotus perniciosus*
1964	-	Egg parasitoid *Telenomus* sp. imported from New Guinea to control Castor semilooper *Achaea janata*
1965	-	Predator *Platymeris laevicollis* introduced from Zanzibar to control coconut Rhinoceros beetle, *Oryctes rhinoceros*

Strategies of Biological Control

There are three strategies in biological control

(A) Classical Biological Control

- ☆ Classical biological control involves the introduction of a pest's natural enemies to a new locality/area where they do not previously present (Mahr *et al.*, 2008). This is usually done by government authorities. More than 172 insect species have been successfully controlled by insect CBC agents worldwide as of now.
- ☆ Successful examples of classical biological control include the management of Cottony cushion scale (*Icerya purchasi). I. purchasi* populations have been suppressed in numerous countries with *Rodolia cardinalis.* In India *R. cardinalis* was introduced in 1929 in Tamil Nadu. Further, efforts were made to multiply and distribution of the vedalia beetle (*R. cardinalis)* in Coorg and Bombay provinces during 1946. Due to these efforts the pest has been nearly controlled and checked its spread to other provinces.
- ☆ *Acerophagus papayae*, a parasitic wasp was imported for controlling papaya mealybug in India.

(B) Augmentative Biological Control

It refers to the process of rearing and releasing biological control agents to reduce the pest population (Hoy, 2008). This method can be categorized into two types: Inoculative releases and Inundative releases

i. **Inoculative releases:** Inoculation is a preventive approach where small numbers of a natural enemy are released early in the pest cycle to prevent the pest population buildup. Pest control is expected from off springs and subsequent generations of released biocontrol agents. It provides long-term control and suppress pest populations from reaching excessive numbers. Ideally, the natural enemies reproduce and build in numbers along with the pest population to provide continued control throughout the season (Toscano, 2017).

Some of the examples of inoculative approach are:

Periodic releases of the parasitoid, *Encarsia formosa,* are used to control greenhouse whitefly, and the predatory mite *Phytoseiulus persimilis* is used for control of the two-spotted spider mite.

ii. **Inundative releases:** Inundation involves the mass culture and release of large numbers of natural enemies to provide a quick, knockdown effect on the pest population. During pest outbreaks, this method is effective as control of pest is expected from released ones. The use of microbial pesticides such as Bt, *Bacillus thuringiensis*, is also a form of inundative biological control (Toscano, 2017).

Some of the examples of inundative approach are:

- Mass release of *Bracon brevicornis* and *Goniozus nephantidis* to control Coconut blackheaded caterpillar, *Opisina arenosella.*

- Inundative release of Green lace wing, lady beetles to control aphids
- Mass release of *Trichogramma* egg parasitoids to control the lepidopteran pests
- Release of *Apertochrysa astur* to control Rugose spiraling whitefly in coconut

(C) Conservation Biological control

Conservation biological control is the protection and conservation of natural enemies against adverse effects of pesticides and incompatible cultural practices and improving their efficiency via providing suitable environment. Natural enemies are already adapted to the habitat and to the target pest, and their conservation can be simple and cost-effective (El-Wakeil *et al.*, 2017).

This can be carried out by:

1. **Habitat manipulation:** Cropping systems can be modified to favor the natural enemies. e.g. Provision of shelterbelt, hedgerow, or beetle bank
2. **Maintaining Diversity:** Planting of Nectar-rich plants that bloom for long periods are especially good, as many beneficials are nectivorous during the adult stage, but parasitic or predatory as larvae.
   i. e.g. soldier beetles (Blister beetle) which is frequently found on flowers as an adult, but whose larvae eat aphids, caterpillars, grasshopper eggs, and other beetles.
   ii. Planting of prune trees in grape vineyards to provide an improved overwintering habitat or refuge for a key grape pest parasitoid.
3. **Provision of artificial shelters of wooden caskets, boxes in gardens:** To make a cropped area more attractive to natural enemies. e.g. the stimulation of the natural predator Dermaptera is done in gardens by hanging upside-down flowerpots filled with straw or wood wool.
4. **Natural enemies inactive Stages Preservation:** keeping the trash in the field itself after harvesting e.g. Pupae of *Epipyrops* (parasite moth) found in the trash of sugarcane leaves at the time of harvesting are left around harvested fields to augment the supply of natural enemies in the pre-monsoon season against Sugarcane pyrilla.
5. **Protection from Pesticides:** Use selective pesticides and target them to pest-problem spots for the protection of beneficial insects
6. **Soil health maintenance:** Provision of adequate organic matter in soil to ensure good habitat for soil microorganisms that control a variety of pests, particularly soil pathogens and root feeding nematodes.

Although each of these biological control methods may not always be successful on its own, when combined, they can be a potent IPM tool.

Biological Control Agents: Biological Control Agents or Biocontrol Agents are defined as those organisms found naturally or genetically engineered that would eliminate or control the pest population to prevent economic losses occurring due to harmful effects from the targeted pests (Mishra *et al.*, 2020). Bio control Agents can be also divided into predators, parasites and parasitoids, and pathogens depending on the organisms used as agents and their affecting mechanisms on pests.

☆ **Predators:** Predators are free living organisms and feeds on several preys to complete its life cycle. They tend to feed on preys smaller than themselves.

e.g: Coccinellid beetles feed on aphids, scale insects, mealy bugs and mites. Both grubs and adults are voracious feeders. *Rodolia cardinalis* is used for the control of cottony cushion scale on citrus. *Cryptolaemus montrouzieri* is commonly used for the management of mealy bugs and scale insects.

Chrysoperla carnea feed on aphids, red mites, thrips, white and black flies, eggs of leaf hoppers, moths, leaf miners and small caterpillars.

Predatory mites, *Phytoseiulus* spp. *Amblyseius* spp. are important in controlling phytophagous mites in several crop ecosystems.

Syrphid, *Ischiodon scutellaris* is used in large scale in mustard and other crops.

☆ **Parasitoids:** Parasitoids are of the same size as the hosts, or sometimes even smaller. Each parasitoid requires only one host, which it kills for its development into a free living adult. The adult food of parasitoids is different from their immature stages. Many adult parasitoids feed on nectar or pollen based on Stage of host attacked the Parasitoids are of different types

a) **Egg Parasitoid:** The egg parasitoids are those individuals that deposit their eggs in and whose progeny emerge from the host eggs.

   e.g. *Trichogramma* egg parasitoids are used the management of tissues borers in sugarcane, stem borer in rice, boll worms in cotton and pests of several crops. *Trichogramma chilonis* is used for the control of bollworms in cotton, intermodal borer in sugarcane and rice leaf folder. *Telenomus remus* is used for the control of tobacco caterpillar.

b) **Egg larval parasitoid:** The egg larval parasitoids are those which deposits eggs in the host eggs but development and emergence are not completed until the host larval stage is reached e.g: *Chelonus blackburni* is used for the control of spotted boll worm.

c) **Larval parasitoid:** The larval parasitoids are those parasitoids which deposit egg on the larvae and their progeny complete development and emerge from the host larvae.

   E.g.: *Campolestis chloridae* is used the control of *Helicoverpa armigera.*

Bracon hebetor and *Bracon brevicornis* for the control of coconut black headed caterpillar.

d) **Larval pupal parasitoids:** The parasitoids those deposit their eggs on or near host larvae and emerge from host pupae.

   e.g. *Isotima javensis* is used for the control of top shoot borer of sugarcane.

e) **Pupal parasitods:** Parasitoids that deposit their eggs in the host pupae and emerge from the host pupae. e.g. *Trichospilus pupivora* and *Brachymeria nephantidis* are used for the control of coconut black headed caterpillar.

f) **Nymphal and adult parasitoids:** The parasitoids deposit eggs on nymphs and their progeny complete development and emerge from the host nymphs.

   e.g. Aphelinus mali is used for the control of apply wooly aphid.

   Encarsia formosa for the control of whitefly.

Entomopathogens

Entomopathogens are disease causing organisms such as bacteria, viruses, fungi and protozoa in insect pest, which kill their host or debilitate the future generations. The infected insects are unable to feed properly, remain stunted, lose their body colour and gets immobile or paralyzed. Under certain conditions they cause disease epizootics in the field.

The various entomophatogens are a) Entomopathogenic bacteria b) Baculovirus c) Entomopathogenic fungi d) Protozoans e) Entomopathogenic nematodes (EPN)

a) **Entomopathogenic bacteria:** These are extra cellular associated with insects and enter the body cavity through contaminated feed and multiply. Death of the insect may be either due to sudden lack of oxygen, intoxication, chemical changes in the gut or by the toxins or crystal produced in the bacterial cells. Several commercial products based on *Bacillus thuringiensis* and its sub sp. is available and widely used for controlling lepidopteran pests (Tanada and Kaya,1993).

   e.g. *Bacillus thuringiensis* var. *kurstaki* against Lepidoptera.

   Bacillus thuringiensis var. *galleraie* against wax moth.

   Bacillus thuringiensis var. *tenebrionsis* against beetles and weevils.

   Paenibacillus popilliae is used against several species of white grubs

b) **Baculoviruses:** Viruses are obligate, sub microscopic, intracellular pathogenic entities. Viruses in the family of baculoviridae are the best known of all the insect viruses Baculoviruses are double stranded DNA viruses having bacilliform or rod shaped virions (Miller, 1997).

   Disease symptoms are easily recognized. Important sub groups within the family are Nuclear Polyhedrosis Viruses (NPV) and the

Granulosis Viruses (GV) which are widely used in pest control the NPVs are specific to host species or genus. GVs are more specific than the NPVs and are reported from Lepidoptera.

Ha-NPV is used for the control of *Helicoverpa armigera* on cotton, chickpea and groundnut. SI-NPV is used for the control of *Spodoptera litura* on tobacco, groundnut, black gram and cotton.

e.g. GV of *Chilo infuscutellus* is used for the management of shoot borer on sugarcane.

C) **Entomopathogenic fungi:** Different enotmofungal pathogens such as white muscardine, green muscardine and yellow muscardine are known to attack various pests. The spores of the fungus directly penetrate integument of pest and body cavity is attack. The fungus with its mycelium and spores cover the body of host (McCoy et al., 1988).

Beauveria bassiana against rice hispa, *Spodoptera litura* and lepidopteran tissue borers sugarcane.

e.g. *Metarrhizium anisopliae* against white grubs and on rhinoceros beetle.

Nomurae rileyi against Tobacco caterpillar and *Hirsutella thompsoni* on mites.

d) **Protozoans:** Protozoa kill the insect either directly or by reducing the fecundity of the adult. They are always associated with other pathogens. Their effect is chronic in nature on host. These are called debilitating infections. They prolong the larval life of the insect pests in the field leads to exposure for longer period to predators and parasitoids.

e.g. Nosema melolonthae against chaffer beetles *Farinocystis triboli* against red flour beetle *Nosema locustae* against grasshoppers

e) **Entomopathogenic Nematodes (EPN): EPNs** are small worm-like animals, found naturally in soils. EPNs from two genera, *Steinernema* and *Heterorhabditis* are used globally to control major insect pests. The invasive larvae of EPNs carry the associated bacteria *Xenorhabdus* sp. or *Photorhabdus* sp. In their gut which not only kill the invaded host within 24-48 hr but also produce a biostatic substance that retards the purification of the cadaver (Rao, C. S. 2011).

e.x. Steinernema carpocapsae and *Heterorhabdits bacteriophora* against soil inhabiting insects and tissue borers.

Merits of Biological Control

- ☆ Specific in action.
- ☆ Safe to non target organisms, environment and to the person who applies them.

- ☆ High level of pest control at low cost and provides long term control with high benefit cost ratio.
- ☆ General inability of pests to develop resistance to bio control agents.
- ☆ Bio control agents give protection to the crop throughout the crop period.
- ☆ Do not cause toxicity to the plants.
- ☆ Chemical residue-free products from farms and natural systems.
- ☆ They multiply easily in the soil and enhance the soil health by encouraging beneficial soil microflora resulting higher crop yields.
- ☆ Bio control agents are very easy to handle and apply to the target.
- ☆ Bio control agent can be combined with bio-fertilizers.
- ☆ It helps in protection of biodiversity and restoring natural ecosystems.

Limitations of Biological Control

- ☆ The process of setting up a biological control system is a costly endeavor. it requires lot of planning, trained man power and money goes into developing a successful system.
- ☆ It's a slow process. The biological agents must be used over an extended period of time in order to have any effect on a pest population; in contrast to insecticides which operate quickly.
- ☆ Subsequent pesticide use restricted.
- ☆ Expensive to culture many natural enemies.
- ☆ Biological control can never eradicate a pest organism completely, because if the control agent reduces the pest population too far, it destroys its own food source.

Conclusion

Growing demand for IPM practices in a number of crops has arisen because of a realization of the problems created by misuse of agrochemicals. Horticultural products free from pollutants are also demanded in the majority of industrialized countries; therefore, management of pests and diseases through biological control is becoming essential. Due to the economic and environmental advantages of biological control, its improvement and expansion of use will certainly be seen in the future. Hence, studies on the identification of potential parasitoids and predators and development of bio pesticides are the potential to undoubtedly revolutionize agricultural practices in the context of their evolution toward being less harmful to the environment. Specified techniques underlying these approaches are developing at constant and are modified to meet the switching requirements of pest management. Modifications and improvements in rearing and application techniques and genetic advancement of natural enemies have increased the effectiveness of biological control agents. Further, application of new ecological theory is transforming the research need for conservation of natural enemies. For its successful implementation with full potentiality, continued refinement and adaptation of approaches and applications are necessary.

References

Carson, R., Darling, L. and Darling, L. 1962. Silent Spring, *Boston:Houghton Mifflin Company Riverside Press (Cambridge Mass)*, pp 61-62.

DeBach, P. 1964. Biological Control of Insect Pests and Weeds. *Cambridge University Press, Cambridge, UK,* pp 844.

El-Wakeil, N., Saleh, M., Gaafar, N. and Elbehery, H. 2017. Conservation Biological Control Practices. *In Biological Control of Pest and Vector Insects,* April. https://doi.org/10.5772/66312.

Hoy, M. A. 2008. Augmentative Biological Control. *In: Capinera, J.L. (eds) Encyclopedia of Entomology. Springer.* pp 327–334.

Johnson, M., Wratten, S. D., Landis, D. A. and Gurr, G. M. 2008. Recent advances in conservation biological control of arthropods by arthropods. *Biological Control* **45**: 172–175.

Mahr, D., Whitaker, P. and Rigdway, N. 2008. Biological control of insects and mites: An introduction to beneficial natural enemies and their use in pest management. *Cooperative extension publishing, Madison, WI.* pp 59.

Stern, V. M., Smith, R. F., Van den Bosch, R. and Hagen, K. S. 1959. *Hilgardia*, **29**, 81-101.

Tanada, Y. and Kaya, H. K. 1993. "Insect Pathology". *Academic Press, San Diego.* . pp 554-594.

Toscano, K. 2017. Augmentation Biological Control Practices for the Home Landscape. *Oklahoma Cooperative Extension Service.* https://extension.okstate.edu/fact-sheets/augmentation-biological-control-practices-for-the-home.

McCoy, C.W., Samson, R. A. and Boucias, D. G. 1988. Entomogenous Fungi. In: Handbook of Natural Pesticides, Vol. V, Microbial Insecticides, Part A, *CRC Press, Boca Raton,* pp 151-236.

Miller, L. K. 1997. "Introduction to the baculoviruses". The Baculoviruses, *Plenum Press, New York.* pp 1-6.

Mishra, P., Tripathi, A., Dikshit, A. and Pandey, A. 2020. Insecticides Derived from Natural Products: Diversity and Potential Applications. In Natural Bioactive Products in Sustainable Agriculture. *Springer Nature Singapore Pte Ltd.* pp 83-99.

Rao, C. S. 2011. Insect Ecology and Integrated Pest Management (Ento-231) - Notes.

Chapter-2

PRODUCTION PROTOCOLS OF ECONOMICALLY IMPORTANT BIO AGENTS

N.B.V. Chalapathi Rao[1], *P. Sunitha*[2] *and V. Anoosha*[1]

[1]*Dr YSRHU-Horticultural Research Station*

Dr. YSR Horticultural University

Ambajipeta, Dr. B. R. Ambedkar Konaseema District, A.P.

[2]*Dr YSRHU-Horticultural Research Station*

Dr. YSR Horticultural University

Venkataramannagudem, West Godavari District, A.P.

Introduction

Chemical pesticides have been in use for the control of pests though they are effective, dependable, economical their indiscriminate use has resulted in several problems such as development of resistance in pests against pesticides, resurgence of minor pests, toxic residues in food, water, air and soil elimination of natural enemies, disruption of ecosystem and their continued uses may further harm the environment and reduce the crop yield.

One of the main obstacle to increase productivity and improving profitability in any agri/ horti production system is the insect pest problem. Pesticide use has undoubtedly helped to protect crops and reduce output losses. During the past three to four decades, the usage of pesticides to manage hazardous pests such as insects, nematodes, diseases, weeds *etc.* has multiplied. Over the time, indiscriminate and unscientific use of pesticides has negatively impacted the ecological balance, leading to pest resurgence, the emergence of pest species with resistance and environmental degradation. The demand for minimising the use of chemical pesticides in the control of insect pests has been driven by growing public concern over the potential health risks of synthetic pesticides as well as significant increases in cultivation costs that result in low profit margins for farmers. There is a need for alternative crop protection strategies due to the numerous drawbacks of using harmful pesticides and their non-judicious use in agriculture. Biological control of pests is the eco-friendly pest management practices useful in almost all agro-ecological situations.

Mass production of parasites and predators is often necessary or desirable in connection with many biological control projects, particularly where attempts are made to increase parasitism or predation by mass releases of bioagents over a wide area at a time in the season when these natural enemies are few or absent. Mass production whether of parasites or predators, fungi, protozoa, bacteria, viruses, sterile males or genetically incompatible individuals has become an integral part of biological control programmes. Mass-production methods usually produce large numbers of entomophages, free from contaminants which can be released immediately in the field. In mass production schemes it is usually easy to eliminate hyperparasites in the case of parasites, or parasites in the case of predators. Diseases and mites, however, may often cause losses in crowded mass-production conditions, and if this occurs rearing methods must obviously be suitably altered. Diseases transmitted through the eggs are particularly difficult to eradicate. (Simmonds, 1964).

Production Protocols of Important Bioagents and Factitious Hosts

The rice meal moth, *Corcyra cephalonica* Stainton (Lepidoptera:Pyralidae) ranks first in the mass culturing of entomophagous insects due to its amenability to mass production, adaptability to varied rearing conditions and its positive influence on the progeny of the natural enemies. *Corcyra cephalonica* (Stainton), a stored grain pest has been proved to be one of the most efficient surrogate hosts for rearing a wide range of biological control agents. The important among them are egg parasitoids - *Trichogramma* spp., egg larval parasitoids – *Chelonus blackburni*, larval parasitoids – *Bracon* spp., *Goniozus nephantidis, Apaneteles angaleti*, insect predators – *Chrysoperla zastrowii sillemi, Mallanda boniensis* are reared on the larvae of *C. cephalonica*. Besides, some entomopathogenic nematodes such as *Steinernema feltiae* is also reared on the larvae of *C. cephalonica*. Only an efficient and healthy insect mass rearing medium can result in mass production of effective biological control agents. *Corcyra cephalonica* can be mass multiplied throughout the year in all the ecological zones of India at 28°C ± 2°C and 65 ± 5% relative humidity considering the economics as well as quality of eggs produced. (http://ppqs.gov.in)

Establishments of Bio Control Units for Mass Production of Bio Agents

Factitious host *Corcyra Cephalonoia* Rearing

1. Preparation of Rearing Trays/Boxes

Wash wooden boxes (45 x 30x 25cm) or rearing drum (40 x 43cm) used for *Corcyra* multiplication with 0.5% detergent and rinsing in tap water followed by wiping with dry, clean – used towel and shade drying. Repeat this process after every rearing cycle and whenever the trays are emptied, they have to be cleaned with 2 per cent formaldehyde and stored.

2. Preparation of larval diet medium for *Corcyra*

a. Sorghum grains are coarsely milled, broken into 2-3 pieces and are heat sterilized at 100°C for 1 hour to decontaminate from other storage pests *viz., Rhizopertha dominica, Sitotroga cerealella, Tribolium castaneum* and fungal contaminants. After cooling these grains were transferred to plastic basins @ 2.5 kg/basin.

b. Blend required quantity of groundnut kernel in a pounding machine or a mechanical blender (mixer grinder) after which 100g of the broken kernel was transferred to each basin and the contents are hand mixed thoroughly.

c. Add dry yeast (Bakers) and wettable sulfur @ 5g/ basin and the contents are mixed thoroughly. Streptomycin sulfate @ 0.05g was also added. This medium is used for rearing Corcyra larvae. (http://agritech.tnau.ac.in)

3. Preparation of *Corcyra* Eggs

a. *Corcyra* eggs used for larval multiplication were maintained free from contaminants like the moth scales and broken moth body parts and not exposed to UV light.

b. The collections of overnight laid eggs are measured volumetrically to ascertain the number of trays that can be infested with eggs. A cc of eggs is known to contain approximately 16000 – 18000 eggs.

4. Infestation of Food Medium with *Corcyra*

Sprinkle 0.25 cc of *Corcyra* eggs on the 2.5 kg sorghum medium which is then covered with lid or a khadi cloth. These basins are carefully transferred to the racks.

5. Handling the Trays During Larval Development

Larvae hatch out in 3-4 days begin to feed on food medium. This can be observed through light webbings on the surface of the food medium. As the larvae grow up they move down into the food medium. The larvae are allowed to grow undisturbed in the trays.

Handling of Adults

Adults begin to emerge in 28-30 days after infestation of the eggs and the adults can be seen on the inner side of the khadi cloth. These can be collected using a test tube (small scale production) or through vacuum cleaner (large scale production). While collecting the adults, the adult rearing room are provided with mosquito nets fitted to the windows and doors to prevent the escape of the moths emerged. Lab personnel are provided with face masks so as to avoid inhalation of scales. Further, these adults are given cotton swab containing adult feed which was prepared by mixing 50 ml honey with 50 ml water and 5 capsules (1 g) of vitamin-E. This feed can also be stored in refrigerator for later use. Piece of cotton wool tied with a thread is soaked in the solution and inserted into the drum through the slot at the top. Moths can be collected up to 90 days after which the number of moths emerging dwindles down and keeping the basins is not economical (http://agritech.tnau.ac.in).

6. *Corcyra* Egg Handling

Large number of eggs are laid by moths are collected on paper at the bottom of the rearing cage. This eggs will be contaminated with scales and broken legs of the moths. And be subjected to sieving after which the cleaned eggs are kept for further use. Daily morning the ovipositor drums are lifted up and the wire-mesh bottoms are cleaned gently with a shoe brush so that the eggs and remnants of scales and limbs settled on the mesh are collected along with those on the filter paper. The eggs are quantified in measuring cylinders and used for building up the stock of *Corcyra* and utilised as per the natural enemy production requirement .

Precautions

- ☆ The oviposition drums are placed on sheets of filter paper in enamel trays which trap effectively the scales that are freely floating in air.
- ☆ Keep oviposition drums in ventilated place near an exhaust fan to enable the workers comfort.
- ☆ Use face masks for the workers to prevent inhalation of scales

Threats in *Corcyra* Production

- ☆ Bacterial disease some time attacks *Corcyra* culture. To control this, streptomycin sulphate be added to the crushed sorghum @ 0.23gm/kg and mixed thoroughly.
- ☆ Occasionally the mite *Pyemotes ventricosus* (Newport) contaminate the culture and affect egg laying and larval development of *Corcyra*.
- ☆ In case of infestation the racks, cages, boxes *etc*. should be disinfected with formalin and placed in the sun for six hours.
- ☆ Boxes containing developing larvae should be dusted with sulphur so that a thin layer of sulphur is present over the sorghum.

- In case of severe infestation, acaricide can be applied. For this, muslin sheets are dipped in a 0.05% solution of acaricide and air dried for couple of hours.
- The sheets are spread over the sorghum grains in boxes. Mites coming in contact with the treated cloth are killed rapidly.
- Control of *Bracon hebetor* infestation: A plastic tub filled with water is placed near racks containing *Corcyra* boxes and keeps a table lamp with 60 or100 W bulb. The light of the table lamp should face the surface of water. Parasitoids will be attracted towards light and in turn will fall into the water and die.
- Another method to trap them is a fine film of some sticky material on the glass window pane and directing the light of table lamp towards glass pane.
- The trap should be operated daily and side by side the culture should also be replaced with fresh stock.
- The boxes should be sterilized by spraying 1% formaldehyde or sodium hypochlorite solution and sun dried for 6 hours. Care should be taken to remove host/parasitoid cocoons from boxes.
- The feed material jowar should also be sterilized. Mix all the ingredients in separate room and charge with *Corcyra* eggs. Close the lid and these sterilized boxes should not be open for 40 days and kept in separate room. By this method, *Bracon* can be eliminated from the culture.
- In case of labs having separate *Corcyra* rearing rooms, on one corner disinfect all boxes one by one and keep them in one room. All the *Bracons* in previous room can be killed by spraying low concentration of any pesticide. All the windows and doors can be kept open for 15 days.
- Thereafter, sprinkle whole room with 1% formaldehyde (40%) solution. After about 10 days bring back whole culture to same room. Thereafter, regularly keep light-water trap to monitor *Bracon*.
- If we find discoloured larvae and white cocoons, then they are collected and immersed in 0.5% formaldehyde to kill the parasitoids.
- Provide windows with fine iron mesh, wire mesh and double door to avoid any entry of parasitoids. (https://niphm.gov.in/)

Mass Production of Parasitoids and Predators

Parasitoids

Production of Trichogrammatids

Trichogrammatids are an important group of bio agents which have been used successfully for biological suppression of various lepidopteran pests damaging crops like rice, sugarcane, cotton, vegetables, maize etc. In India, *Trichogramma* spp., are in great demand today and their mass production has already been taken up by various cooperatives, sugar factories and other agencies in public and private sector. Trichogrammatids can be reared very easily using glass tubes (8" x 1.3"). Egg cards made of about 10,000 frozen eggs of *Corcyra* are exposed to the adult female parasitoids(*Trichogramma* spp.) in the ratio of 20:1. After exposing for 24 hrs, the egg card is removed and 1/3 of it can be cut and transferred to another tube to be used for maintain pure culture. The remaining parasitized egg cards can be utilized as nucleus culture for shipment or field release. The adult *Trichogramma* emerge from the parasitized card 6 -8 days later. This method can be used for rearing all *Trichogramma* species.(http://www.aau.ac.in)

Bracon hebetor/ B. brevicornis

It is widely distributed and recorded from *Antigastra catalaunalis, Adisura atkinsoni, Earias* sp., *Helicoverpa armigera, Noorda moringae, Opisina arenosella, Pectinophora gossypiella, Chilo partellus* etc.,. Though Its natural parasitism is not very high, it paralyses many lepidopteran larvae and some of them never revive. This contributes to indirect control.(https://agritech.tnau.ac.in/) These parasitoids were mass multiplied in the laboratory on the alternate host *Corcyra cephalonica*. Grown up caterpillars were spread on stretched cloth over the mouth of jar and a glass plate is placed over the larvae to immobilize. Adult parasitoids were released in the jar. The female parasitoid paralyses the host before depositing its eggs on the body of the caterpillar. The eggs hatch out into grubs and feed on the body of the caterpillar. The grubs before spinning cocoons were collected on to a clean piece of paper (2"x1"). This piece of paper is sufficient for grubs to spin cocoons and attach themselves to the paper. Emergence of parasitoids start from 7th day of parasitisation. The adult parasitoids were released in the black headed caterpillar infested garden at the third instar larval stage of the host.

Goniozus nephantidis

Goniozus nephantidis is the most widely used parasitoid of *Opisina arenosella*. It is a sturdy gregarious larval or prepupal ectoparasitoid. The female practices maternal care of eggs and larvae. The host larvae are parasitized and the parasitoid even feeds on host body fluid. The parasitoid is also capable of suppressing the population by merely stinging and paralyzing 1st – 2nd instar *larvae. G. nephantidis* is the most common and effective parasitoid of late instars caterpillars of *O. arenosella* in several parts of the country.This parasitoids were reared on the alternate host, *Corcyra cephalonica*. Parasitoids on emergence were placed together in tube (6"x1") for mating and fed on honey solution. After 24 hours, the mated female is separated into a specimen tube (3"x1") and a single well developed caterpillar introduced. The parasitoid at first paralyses the caterpillar by stinging it once or twice after which it starts oviposition. In this case also the grubs were removed on to strips of paper (2"x1"). After oviposition, the female parasitoid were removed, mated and fed on honey solution and used once again after 24 hours, for setting in a new cage. This process was repeated until such time the parasitoids die. (https://agritech.tnau.ac.in/)

Predators

Production of Predacious Coccinellids

Coccinellids prefer to feed on aphids, mealybugs, white flies but when these preys are not available they switch to other prey such as mites, thrips, lepidopteran eggs etc. both adult and larvae are voracious feeders. During its development through four instars, the larvae eat between 200-300 aphids and adult may eat up to 50 aphids per day.

Coccinelids for Laboratory Rearing

A plastic container with a capacity of 1-2 ltr. has to be taken and a piece of cotton dipped in water and another one dipped in 10-15% honey solution in a petri plate has to be kept inside the container.A container (small cup) with mustard seedling/cowpea infested with aphids are placed in the container and 5-10 adults of coccinellids (mixed sex) are released in each container. A corrugated white paper is placed inside the container for egg laying by the coccinellids. The eggs laid by the females will be collected daily by cutting of the paper with eggsand transferred to a petri plate. The eggs will be incubated in the growth chamber at 28+6°C with 65% RH. On hatching the grubs may be transferred into new container with similar setup as described earlier (http://www.aau.ac.in).

Production of Predacious Reduviids

The Reduviids, also known as assassin bugs, are of considerable economic importance because they reduce the pest population by killing the host quickly with their proteolytic saliva. The effectiveness of the reduviids as biocontrol agents has been demonstrated and the field release usually resulted in quick and effective control of the target pests. Reduviids can be reared on natural or alternative hosts. In the laboratory, reduviids are mass reared using *Corcyra cephalonica* larvae as host.

Reduviids for laboratory rearing

A plastic tub is taken with sterilized sand to a depth of 2 cm. A cotton swab dipped in 10 -50 % honey solution is placed on the side wall of the tub and corrugated white paper for oviposition is attached inside the plastic tub. About 2-3 grown up larvae (4th-5th instars) of *Corcyra* or natural hosts (*Helicoverpa/Spodoptera* or any other lepidopteran host caterpillars are added inside the plastic tub and 2-3 pairs of young adults (male and female in 1:1 ratio) of reduviids are released inside the plastic tub. The open mouth of the plastic tub is covered with a muslin cloth immediately after releasing the reduviids. The cotton with honey solution is to be changed every two days. *Larvae* of lepidopteran host are to be added every two days. The female reduviids lay eggs on corrugated white paper in clusters of 40-50 eggs. The egg masses from the white paper are collected by cutting out the paper with eggs on it. The paper with egg masses should be placed in a petri dishes and cover it with a lid. Egg hatch into nymphs in 7-10 days. The nymphs are maintain in petri dishes for a week by adding 3-4, $1^{st}/2^{nd}/3^{rd}$ instar host larvae or as needed. The nymphs are transferred after a week to a plastic container or tub (10 -20 / container) with sand (2cm thickness) at the bottom with a piece of corrugated white paper. A cotton swab dipped in 10-50 % honey solution on the side of the wall is to be placed. The eggs laid by the reduviids may be transferred to new container again with similar steps as described (http://www.aau.ac.in).

Production of Neuropteran Predator *Chrysoperla zastrowi sillemi*

The green lacewings, *Chrysoperla* (Neuroptera: Chrysopidae) is a cosmopolitan predator found in a wide range of agricultural habitats. They are important for the management of bollworms and aphids in cotton and tobacco and several sucking pests in fruit crops. They are capable of bringing down the population of the pest drastically. In India, 65 species of chrysopids belonging to 21 genera have been recorded from various crop ecosystems. Some species are distributed widely and are important natural enemies for aphids and other soft bodied insects. Amongst them, the green lacewing *Chrysoperla zastrowi sillemi* is the most common. *Chrysoperla zastrowi sillemi* is now used extensively all over the country. It is being mass produced primarily on the eggs of rice grain moth, *Corcyra cephalonica* in India. In mass production the adults are fed on various types of diets. The larvae are reared either in plastic tubes or empty injection vials.

The adults are collected daily and transferred to plastic vial or glass troughs (30 cm × 12cm) before allowing the adults, the rearing troughs are wrapped inside with brown sheet which act as egg receiving card. About 250 adults (60% females) are allowed into each trough and covered with white nylon or georgette cloth secured by rubber band. On the cloth outside three bits of foam sponge (2 sq.in) dripped in water is kept. Besides an artificial protein rich diet is provided in semisolid paste in three spots on the cloth outside. The adults lay eggs on the brown sheets. The adults are collected daily an allowed into fresh rearing troughs with fresh food. From the old troughs, the brown paper sheets along with *Chrysopa* eggs are removed. The sheets containing *Chrysopa* eggs are stored at 10°C in BOD incubator or refrigerator for about 21 days. When the eggs are required for culturing or for field release the egg sheets are kept in room temperature for a day and eggs during these peiod turn brown and hatch on 3 days later. The first *larvae* are either taken for culture or for recycling or for field release. (http://ppqs.gov.in ; http://www.aau.ac.in).

Rearing of Neuropteran Predator *Apertochrysa astur* adults

The known (15 adults/ container) numbers of freshly emerged adults were collected into a plastic pet jars (10cm x 15cm). Provide the container with cotton swab dipped in a mixture of honey + protein x + fructose + yeast + water (equal ratio)as adult food and then a brown paper sheet was placed inside the rearing jar lid for harbouring the eggs laid (egg card), make a pin-size holes on sheet for aeration. The brown sheet has to be changed on every alternate day along with fresh food. Collected egg sheets were cut into small clippings by quadrate method for future distribution, future multiplication and storage.

Rearing of Predator *Apertochrysa astur* Grubs

Freshly emerged grubs were fed with UV sterilized eggs of surrogate host *C. cephalonica* were sprinkled @ 0.2 cc sterilized eggs/ 50 predator eggs in plastic pet jars (10cm x 15cm), place finely cut newspaper strips folded at 2cm interval (30 cm x 2 cm) around 10 in numbers inside the jars was done to avoid cannibalism. After two days, the emerged grubs were collected and transferred individually with a fine camel hair brush into a small plastic vial with cap (3cm x 4.5cm) having pin size holes for aeration in which sterilized *C. cephalonica* eggs were also provided to facilitate the larval feeding and were randomly checked in five days interval. Approximately 1 cc of sterilized *C. cephalonica* eggs were required for rearing 20 *A. astur* grubs placed individually in the vials till pupation and in turn these vials were placed in basins which is covered by georgette black cloth secured by rubber bands and marked with date of egg feed provided. Collect the adults after 10th day onwards.

After two days, the emerged grubs were collected and transferred individually with a fine camel hair brush into a small plastic vial with cap (3 cm x 4.5 cm) having pin size holes for aeration in which sterilized *C.cephalonica* eggs were also provided to facilitate the larval feeding and were randomly checked in alternate four to five days. At this stage approximately, 1 cc of sterilized *C. cephalonica* eggs were required for rearing of twenty *A.astur* grubs. The grubs are placed individually in the vials till pupation and these vials were placed in basins which are covered by georgette black cloth secured with wooden frame. To facilitate large scale predator production 70 vials each were placed in each basin which were in turn placed on wooden racks (3.2 m × 0.75 m × 1.5 m) with nine compartments so that each rack can hold 16 basins. Maintenance of sufficient *C.cephalonica* eggs supply was ensured to facilitate the target predators eggs production. The adults emerge from 10th day onwards and emergence continued till 15th day. The emerged adults were carefully transferred with help of test tube into designated clean adult rearing jar (10 cm dia 15 cm height) with a brown paper inside the lid. The brown paper sheet kept inside the rearing jar harbours the eggs laid and when required for culturing or for field supply to the farmers the egg sheets were cut into fine strips with each strip containing 5 to 7 eggs and placed in card board boxes of size 23 x 23 x 9 cm which can accommodate ten thousand eggs for safe transport.

Anthocorid bugs

Anthocorid bugs, particularly many species of the genus *Orius* (Heteroptera: Anthocoridae) are important beneficial insects for various agro systems feeding primarily upon thrips, aphids, leafhoppers, psyllids, spider mites, eggs and young larvae of several crop pests. *Cardiastethus pygmaeus* Poppius, *C. affinis* Poppius and *C. exiguus* Poppius are key predators of coconut leaf caterpillars. *Opisina arenosella* Walker and *Xylocoris (Arrostelus) flavipes* (Reuter) are associated with pest of stored grains / commodities.

Cardiasthethus exiguus: Adults of *Cardiasthethus exiguus* (15 males and 15 females) were released into a cylindrical glass jar 17.5 cm x 6.75 cm, with cotton strands for oviposition as substrate with 300 *corcyra* eggs and pollen dust and cotton swab (water soaked) which is stuck to the wall of container. The egg laying was observed evenly on the walls of containers and these eggs were gently removed to the other container. Fresh cotton and *corcyra* eggs were supplied for every 24 hours. To prevent cannibalism, the eggs were separated from the adults of the predator.

With the increased need and awareness of integrated pest management concept among the farmers, there is increased emphasis on the utilization of biocontrol agents for the management of pests. Though their demand is increasing, yet their availability is far from sufficient. The bio control agents, particularly parasitoids and predators, have short life span and they cannot be stored for long. Their transportation also requires certain specific conditions which are difficult and expensive. These are some of the important reasons that private sectors are not much interested in them.(Kumar Pradyumn, 2017). However, biological control is making deep inroads into the farmers heart due to its long term impact in eco-friendly management of pests.

References

http://www.aau.ac.in/

https://agritech.tnau.ac.in/

https://niphm.gov.in/

Kumar, P., Kaur, J., Sekhar, J. C., Soujanya, P. L. and Suby, S. B. 2017. Mass production of biocontrol agents of insect pests. In: Industrial Entomology, Springer publications. pp 451 – 465.

Simmonds, F. J. 1964. Mass Production of Insect Parasites and Predators. Bull. Org. mond. Sant 31: 511-512

Chapter-3

RESEARCH STRATEGIES TO OVERCOME THE CHALLENGES IN APPLICATION OF BIO AGENTS

K. Sireesha[1], V. Anoosha[2] and G. Shali Raju[3]

[1]Dr YSRHU-Horticultural Research Station
Dr. YSR Horticultural University
Lam, Guntur District, A.P.
[2]Dr YSRHU-Horticultural Research Station
Dr. YSR Horticultural University
Ambajipeta, Dr. B. R. Ambedkar Konaseema District, A.P.
[3]Dr YSRHU-Krishi Vigyan Kendra
Dr. YSR Horticultural University
Venkataramannagudem, West Godavari District, A.P.

Introduction

Though it is evident that biological control programmes have been successfully applied in a variety of crop environments throughout the countries with increasing scope of adoption and potential to expand it is still true that this method of control is slow to be adopted and put into practise. Ecologically sound pest control has not been adopted for a range of pests and diseases due to technical, economic, but primarily psychological barriers. Although funding for biological control research is limited, now a days it is realised that the benefits accrued are of long term in nature and cannot be ignored. It is certain that biological control will play a significant role in pest management in the future because it is the most cost-effective, environmentally safe, and sustainable method of pest control with extra advantages for growers and consumers. Hence there is a need to continuously strive both at national and global level in strengthening the agencies involved in bio control research and promotion and simultaneously identify the bottlenecks and strategies to overcome the same.

Agencies Involved in Bio-Control Research and Promotion International

The International Organisation for Biological Control (IOBC), Zurich, Switzerland

The International Organisation for Biological Control is a global organization with a mandate for the promotion of environmentally safe methods of insect, weed, and disease control. This organisation effectively advocates and promotes biological control and initiates necessary steps to influence the decision makers in this path . Established in 1955, IOBC operates internationally via geographical Regional Sections and topic-focussed working groups (global and regional). The organization promotes national and international research, training of personnel, coordination of large-scale applications, and public awareness of the economic and social importance of biological control. IOBC arranges conferences, meetings, and symposia, and takes other actions to implement the general objectives of the organization (Barratt *et al* 2018).

Membership in IOBC allows individuals and organizations to participate in biological control activities beyond their specific jobs and workplaces, to step outside their bureaucracies, and to contribute to the promotion of biological control worldwide.

The Mission of IOBC is:

- ☆ IOBC promotes the development of biological control and its application in integrated pest management programs, and international cooperation to these ends.
- ☆ IOBC collects, evaluates, and disseminates information about biological control, and promotes national and international action concerning research, training of personnel, coordination of large-scale application, and public awareness of the economic and social importance of biological control.
- ☆ IOBC arranges conferences, meetings, and symposia, and takes other actions to implement the general objectives of the organization.
- ☆ As an independent professional organization, IOBC can be an effective advocate for biological control and can influence policymakers and Governments.
- ☆ IOBC assists in the communication among biological-control workers, through dissemination of the IOBC Newsletter.

☆ IOBC publishes a journal for basic and applied research on biological control of invertebrates, vertebrate and weed pests, and plant diseases. This journal is Bio Control, published by Springer. (IOBC; http://www.iobc-global.org/index.html)

In addition to the Global organization, IOBC has 6 Regional Sections which are

Regional Sections

APRS	:	Asia and the Pacific Regional Section (India, South-East Asia, Japan, and Australia)
ATRS	:	African Tropical Regional Section (African countries south of the Sahara)
EPRS	:	East Palearctic Regional Section (Eastern Europe)
NRS	:	Nearctic Regional Section (North America)
NTRS	:	Neotropical Regional Section (Latin-American countries)
WPRS	:	West Palearctic Regional Section (Europe, Mediterranean Region, Near- and Middle East)

National (within India)

In India, research on biological control of insect pests and weeds was initiated in the early sixties under PL 480 grant. Subsequently, realizing the evils of chemical pest control methods like insecticide resistance, pest resurgence, and destruction of bio-diversity and appreciating the need for research in this non-chemical and safe method for pest management, the Indian Council of Agricultural Research (ICAR), New Delhi, launched an All India Coordinated Research Project (AICRP) on Biological Control of Crop Pests and Weeds in 1977 with full financial support by the Department of Science and Technology, Government of India. Recognition of the importance of biological control came during the VIII plan with the creation of the Biological Control Centre (BCC) which was functioning under the administrative control of NCIPM, Faridabad. A greater thrust for the planned biological control program started in 1987 when ICAR took over the erstwhile Commonwealth Institute of Biological Control (CIBC), its insect collections and physical facilities.

National Bureau of Agricultural Insect Resources (NBAIR), Bengaluru

Formerly National Bureau of Agriculturally Important Insects (NBAII) is located in Bangalore, Hebbal on the same premises at which The Commonwealth Institute of Biological Control (CIBC), Indian Station was established in 1957. The advent of CIBC marked the beginning of organized and systematic biological control research in India. During this period, knowledge of natural enemies of crop pests and weeds increased manifold. CIBC Indian station was closed in 1987 and the All India Coordinated Research Project on Biological Control of Crop Pests and Weeds (AICRP-BC&W), which was launched in 1977 under the aegis of the Indian Council of Agricultural Research was shifted to the same campus in 1988. The center was named as Biological Control Centre and the entire program functioned under the administrative/financial control of the National Centre for Integrated Pest Management (ICAR). In the eighth five-year plan, the project was elevated to an independent Project Directorate of Biological Control, with its headquarters in Bangalore in 1993. PDBC was the nodal agency in the country that organizes biological control research at the national level with 16 centers spread across the country. The Directorate at Bangalore carried out basic research on the biosystematics of important groups of insect bioagents. The reference collection maintained at PDBC was cataloged in the form of a technical bulletin and also available in a retrievable, electronic format. Besides, work on strain development, molecular characterization, mass production technologies, semiochemicals, and biopesticides work for insect and disease management was intensified. During the XIth plan, PDBC was upgraded as the National Bureau of Agriculturally Important Insects (NBAII) to act as a nodal agency for the collection, characterization, documentation, conservation, exchange, and utilization of agriculturally important insect resources (including mites and spiders) for sustainable agriculture. In the twelfth five-year plan the Bureau is now re-named as National Bureau of Agricultural Insect Resources (NBAIR) and the bureau's activities are divided into three divisions.

The AICRP on Biological Control of crop pests and Weeds was initiated with 13 cooperating centers in 1977 to develop eco–friendly biological control methods for the sustainable management of pests under the aegis of the Indian Council of Agricultural Research (ICAR), New Delhi. The headquarters of the AICRP on Biological Control of Crop Pests and Weeds was located initially at IIHR, Bengaluru, and later at the Project Directorate of Biological Control (PDBC) / later upgraded to the National Bureau of Agricultural Important Insects (NBAII)/ National Bureau of Agricultural Insect Resources (NBAIR). At present, NBAIR as the AICRP-BC coordinating cell coordinates the biological control work of our country through 21 SAU-based centers, 10 ICAR institute-based centers, and 2 central university centers located in different agroclimatic zones (https://www.nbair.res.in/)

Directorate of Plant Protection Quarantine and Storage, Faridabad

The Directorate of Plant Protection Quarantine & Storage was established in the year 1946 on the recommendation of the Woodhead Commission as an apex organization for advising the Government of India and state governments on all matters related to Plant Protection. The Directorate is headed by a Plant Protection adviser. Plant Protection strategy and activities have significant importance in the overall crop production programs for sustainable agriculture. Plant protection activities encompass activities aimed at minimizing crop losses due to pests through integrated pest management, plant quarantine, regulation of pesticides, locust warning and control, and training in desert areas besides training and capacity building in plant protection. It is an attached Office of the Ministry of Agriculture and Farmers Welfare. One of the activities of this organization is the Promotion of the use of bio-pesticides and bio-control agents in plant pest management for which State Bio-control Laboratories (SBCLs) in States/UTs were established under Grants-in-Aid by the Government of India during the Xth and XIth Plan

Bio Control Laboratories

Aims and Objectives of the Bio-Control laboratories are

- ☆ Multiplication of bio-agents in the laboratory and subsequent release in the farmer's fields against their target pests.
- ☆ Conservation and augmentation of natural enemies already present in the farmer's fields.
- ☆ Monitoring and surveillance of insect-pest, disease, weeds, and bio-agent situations on major crops.
- ☆ Training to the trainers & farmers in the identification, production, utilization, and evaluation of bio-control agents.
- ☆ Standardization of methods of mass production of predators, parasitoids & pathogens.

Numbers of Bio-control Laboratories in India

Sl. No.	Type of Lab	Number of Labs
1.	CIPMCs	35
2.	SBCL (Grant in aid)	38
3.	ICAR	49
4.	SBCL	98
5.	Private	141
	Total	361

Source: *https://ppqs.gov.in/*
SBCL: State Bio-control Laboratories, ICAR: Indian Council of Agricultural Research, CIPMCs: Central Integrated Pest Management Centres

National Institute of Plant Health Management (NIPHM), Hyderabad

NIPHM is a National level institute under the administrative control of the Department of Agriculture & Cooperation, Ministry of Agriculture and Farmer's Welfare, Government of India established in the year 1966 as Central Plant Protection Institute (CPPTI) at Hyderabad. The Institute was rechristened subsequently as National Plant Protection Training Institute (NPPTI). It became an autonomous body in the year 2008 with the expanded scope of promoting environmentally sustainable Plant Health Management practices in diverse and changing agro-climatic conditions and Plant Biosecurity Management and Pesticide Management through capacity building programs, besides providing inputs for policy formulation on Plant Health Management, Plant Biosecurity, Invasive Alien Species, market access, Pesticide Management, etc. at the state and national level. This institute promotes biological control through training and capacity building programs on production protocol for bioagents and quality assessment and quality management of microbial biopesticides and organizing short-term courses *viz.,* production protocol for biocontrol agents and biopesticides and quality assessment and quality management of microbial biopesticides, providing hands-on practices for quality analysis on quality standards such as testing of *Helicoverpa armigera*, Nuclear Polyhedrosis virus (HaNPV), *Trichoderma*, *Pseudomonas*, Entomopathogenic fungi, etc. (https://niphm.gov.in/)

Biological control of insect pests is characterised by a persistent, strong reduction in the pest population following the introduction of a natural enemy (Beddington *et al* ., 1978).Biological control is observed to be less spectacular than most physical or chemical controls but is usually also more stable and longer lasting (Baker and Cook, 1974). Despite biological controls having been used in agriculture for centuries, as an industry biological control is still in its infancy. Biological control is now being considered for an increasing number of crops and managed ecosystems as the primary method of pest control. One reason for its growing popularity is its record of safety during the past 100 years considered as the era of modern biological control (Waage and Greathead, 1988). Inspite of its vast potential and scope Biological control programs faces many implementation and execution difficulties and a few important are detailed below.

Challenges in Bio-control

Habitual Pesticide Usage

Pesticides are used to increase agricultural productivity, but they are used indiscriminately and are one of the main the reasons for pollution.

India stands 12th in pesticide use globally and 3rd in Asia after China and Turkey. Only around 0.1 percent of pesticides are believed to reach the intended organisms, with the rest polluting the environment and causing environmental harm Carriger *et al.* (2006); Gill and Garg (2014). There are 293 pesticides registered in India, and it is reported that 104 pesticides are still being produced/used in the country despite being prohibited in two or more nations around the world (GOI 2021). Pesticides may also have more subtle effects on the physiology of natural enemies than direct toxicity. Several fungicides, such as benomyl, thiophanate-methyl, and carbendazim, inhibit oviposition by predacious phytoseiid mites. Certain herbicides (diquat and paraquat) make the treated soil in vineyards repellent to predacious mites.

Despite several advantages like environmental friendliness, selectivity, ability to self-perpetuate, no/low build-up of resistance along with the low cost of deployment, the biological control of insect pests could not compete with chemical control of insect pests as many of the farmers are habituated to use chemicals due its ease of availability and quick action which can be possibly due to the wide publicity and promotion by the corporates involved in pesticides business.

Impact of Climate Change on Bio control Agents

Climate change has impact plant productivity which can be attributed to changes in pests and diseases. (Batalden *et al.*, 2007). The fitness of natural enemies can be altered in response to changes in herbivore quality and size induced by temperature and CO_2 effects on plants. The susceptibility of herbivores to predation and parasitism could be decreased through the production of additional plant foliage or altered timing of herbivore life cycles in response to plant phenological changes.(Thompson *et al.*, 2010). As the performance of macrobials and microbials is highly linked to the climatic actors / environmental conditions and with the present scenario of global climatic change there can be a tremendous impact on the performance of bio agents and consequent farmers' faith too.

High-value crops

The utility of biological control is limited in high value crops as some of these flower and fruit crops that are grown almost exclusively for aesthetic value and export markets must be of highest quality and should have zero pest status and any incidental insect pests or the presence of natural enemies may cause the importers to reject the consignments.

A large number of crop groups and cultivars

In the era of mixed cropping and intercropping being promoted for doubling the farmer's income multiple / inter and diversified cropping systems are being promoted As natural enemies may each have specific hosts it will be highly difficult to adopt biological control for the entire cropping scenario. The multitude of pests in the cropping systems if planned to be controlled with bio-agents create a greater difficulty for the farmers as several different species of natural enemies must be released simultaneously are sequentially thereby enhancing the cost of plant protection as compared to the usage of single broad-spectrum insecticide which can be effective against all insect pests .

Intraguild competition

Intraguild predation occurs among natural enemies in biological control systems, where one natural enemy (the intraguild predator) attacks another species of natural enemy (the intraguild prey), whereas they also compete for the same pest. Intraguild predation is a combination of exploitative competition and predation among potential competitors that use the same host or prey (Arim *et al.*, 2004; Polis *et al .*, 1989) Intraguild predation may occur among only predators, only parasitoids, or different combinations of parasitoids and predators.

This intraguild predation has the potential to greatly limit biological control. Multiple natural enemies may interfere with one another through competition and/ or intraguild predation. For example, Coccinellids struggle with another group of insects outside the family for food resources. They usually co-occur with chrysopid (Neuroptera: Chrysopidae) larvae and share limited resources. However, IGP interactions between them are asymmetrical which could have positive, negative, or neutral impacts on aphid biocontrol *Harmonia axyridis* mostly acts as an intraguild predator and declines local populations of chrysopids and predatory gall midge along with that of aphids (Gardiner and Landis, 2007).The field studies by Ankita Gupta *et al.* 2021 on cassava mealybug (CMB) *Phenacoccus manihoti* revealed that many indigenous parasitoid species actively parasitized the potential native predators of CMB and benefitting the mealy bug. The predators such as *Hyperaspis maindroni* and *Mallada desjardinsi* which are subjected to high parasitism pressure and intra-specific competition from coexisting predators (*Autoba silicula, Spalgis epius,* spiders, etc.).

Narrow host range

Generalized predators may be good natural enemies but they don't kill enough pests when other types of prey are also available thereby providing insufficient control . Generalist predators are widely acknowledged to contribute valuable levels of biological control in agroecosystems throughout the world. Although their feeding habits can result in the rejection of target pests in favor of preferred and often more nutritious non-pest prey, these natural enemies are capable of colonizing habitats prior to the arrival of pests by subsisting on alternative sources of food. The effect of consuming non-pest species on rates of pest predation by a generalist predator can be twofold; feeding upon these nutritious food items generally enhances fecundity thus improving their population growth, but the presence of alternative prey, especially during times when pest regulation is required, can result in reduced levels of pest consumption per individual predator (Harper *et al.*, 2005; Harwood *et al.*, 2004)

Incompatibility with Insecticides

The complex of pest species belonging to various orders attacking a single crop that has specific natural enemies to a particular order poses a challenge to bio control-based pest management However, the usage of insect growth regulators and biopesticides and green label pesticides with short residual toxicity can provide a scope for integration with the use of natural enemies. However research efforts lead to development of Insecticide and temperature tolerant bio agents but very few successful case studies are available.

Taxonomic Constraints

As many parasitoids are species specific correct identification of the pest is crucial in the search for its natural enemy. In case the pest is having sibling species, these sibling species should be correctly identified for effective control. As parasitoids are adapted to the life cycle, physiology and defenses of their hosts, they have a limited host range and are highly specialized. Thus, accurate identification of the host and parasitoid species is critically important in using parasitoids for biological control.

Regulatory Barriers

The use of biological control agents may be subject to regulatory barriers and restrictions in many countries, making it difficult to obtain approval for their use. The guide lines for Guidelines for the export, shipment, import and release of biological control agents and other beneficial organisms needs to be adhered to as per the standard which is intended to facilitate the safe export, shipment, import and release of biological control agents and other beneficial organisms. Responsibilities relating to this are held by contracting parties, NPPOs or other responsible authorities, and by importers and exporters.(International Phytosanitary Portal (IPP) at https://www.ippc.int/coreactivities/standards-setting/ispms.)

Impact of Environmental Conditions on Bio Agents

Temperature is the main environmental factor that influences insect development . High temperatures, short day lengths, or prolonged cool conditions can limit the efficiency of bio agents and can influence their reproduction, and cannot be used at certain times of the year without supplemental light and/or heating.

The adults of many predators and parasitoids do require or benefit from pollen, nectar to effectively reproduce. Many crop plants flower uniformly for only a short time therefore a diversity of various flowering plants along the boarders of the field or within the field may be needed as supplemental sources of pollen and nectar.

Scale of production of Bio agents

The Bio agents particularly macrobials are mainly produced by small and medium-sized companies / Government institutes with a strong research base in biocontrol. There by a small group of producers had to compete with a much larger chemical plant protection market, which results in a very small percentage of market share for BCA in the plant protection sector. While it is cheap in the long run, the process of actually setting up a biological control system is a costly affair

Cost of Production of Bio agents

Biological control agents may be expensive to produce, transport, and release. Moreover, the process of introducing natural enemies into an ecosystem may require a long-term commitment to achieve sustainable control.

Awareness levels in the farmers

The farmers are usually poorly informed about biological control as there are only a few active promoters particularly from the Government sector emphasizing this aspect.

Maintenance and Availability of Host Cultures

The Scale of production adopted by small and medium-sized companies / Government institutes with limited orders and associated manpower poses the challenge of the continuous host insect cultures maintenance and scaling up its production at times of demand. Further insect cultures maintenance requires dedicated and skilled workers round the clock and hence the increase in associated costs too

Perishable in nature

Unlike chemical insecticides the longevity and storage problem is present for microbial bio agents, their life span is limited and field usage depends on the target pest stage requires a high synchronization of production and subsequent stage-based release for which high precision-based decisions by experts are required

Application Techniques

Because bio control agents especially macrobials are released in various stages like egg cards clipping grub release, adult release guidance on proper release techniques has to be imparted to the end user for successful and continuous adoption .

Lack of Active Networking among Bio-Control Workers

To harness the potential effectiveness of bio-control agents, strong bio-control workers are needed to bring partnerships with different disciplines to monitor and evaluate the IPM packages involving bio-control agents for which coordination among research and extension wings is paramount .

Commercial biological control agents are becoming increasingly important in modern, sustainable agriculture. They have gained the attention of developing and emerging country agricultural administrations because of their relatively low toxicity to man and the environment, potential for local production, and compatibility with smallholder farming, which is the predominant form of agricultural production in India. Hence the challenges encountered cab be effectively tackled by adopting the following strategies

Strategies for promoting biological control

The commercial mass production of bio agents and assuring their availability: The preventative establishment of bio-control agents is essential before the arrival of pests and is the key aspect for the success of any bio-control programme hence large scale production by entering MoAs with different private companies for mass multiplication is essential and vital and steps should be planned towards reaching this goal

Periodic releases of bio-control agents: Bio-control agents are unable to continue their life cycle due to climate change or weather conditions or lack of host crops or removal of an entire habitat system. In this situation bio agents may be needed to be re released at regular intervals i.e., augmentative bio control. Where in increased numbers of bio-agents are released into a system along with creating suitable environment for promoting them. Periodic releases may be inoculative releases and innundative releases. Inoculative releases of biocontrol agents are needed to establish and maintain the pest under ETL. Inundative release is needed for long-term balance with the population with its prey to suppress the pest population directly; here the control is from the release itself.

Bio agents production Start-ups: Start up companies have to be promoted and incentives should be planned to sustain them for a certain amount of time to overcome the gestation and initial establishment costs .

Farmer's participation: Farmer understands and participation is crucial at every step of implementation as they are the end users and beneficiaries.

Government support: Pesticide should be priced on a higher scale as compared to bio agents so as to wean the farming community at least initially from the usage of the pesticides and to make biocontrol agents an attractive alternative. There is an immediate need to provide appropriate policy support for the production, promotion, and marketing of quality biopesticides and bio agents. Further, Pesticide industry should develop target specific ecologically safe green labelled molecules which are compatible with bio-control agents.

Communication with farmers: The end users and beneficiaries the farmers are usually poorly informed about biological control, and indeed sometimes misunderstand and disbelieve it as a alternative option for pest management. Hence publishing popular articles in vernacular languages short films and giving improved market access for products produced so can motivate and drive farmers in adopting this ecofriendly technology .Communication and information provided by scientists should be easily comprehensible and in simple language so that farmers can easily understand and adopt. Developing apps for Smartphones, tablets, and computers can help farmers move towards ICT-based Extension (Information Communication Technology) an effective instrument that can reach farmer ing community within a short time.

Promoting Cost-effectiveness of bio-control and integration of supportive players

Promoters of bio control should invariably involve agricultural economists, extension specialists so as to obtain the showcase the social, economic and environmental benefits including pollution control .

Field conservation of bio-control agents

i) **Minimise pesticide usage:** Bio control agents are very sensitive to pesticides, hence judicious usage of pesticides is needed. The use of the right type of insecticide, the right dosage, the right time, and the right method of application in the right direction can play a significant role in minimizing the contact between a natural enemy and the pesticide. (Barratt et al., 2018)

ii) **Selection of pesticides:** Pesticides affect the natural enemies by reducing the longevity, fecundity, and searching ability and decimate the population of flying insects. It can be avoided by the usage of microbial pesticides instead of synthetic pyrethroids and less persistent insecticides can have a less harmful impact on bio-control agents because it will not affect the physiological state of bio-control agents. (Barratt et al., 2018)

iii) **Trap crops:** Growing trap crops along with the main crop increases the overall efficiency of natural enemies.

iv) **Avoiding natural enemy destruction:** Insecticide selectivity: Use of selective chemicals that have a stronger depressive influence on the pest population than on its natural enemies *i.e.* physiological selectivity like usage of microbial insecticides and insect growth regulators.

   Another way is ecological selectivity usage of nonselective chemicals in a selective manner viz., following recommended economic threshold levels, treating with lowest rates possible, timely treatment and usage of selective insecticide formulation, avoiding treating broad areas because untreated areas can serve as refugia for natural enemies, this is best for highly mobile parasitoids like braconids.

v) **Suppression of ants:** Suppression of ants increases the efficiency of predators.

vi) **Insecticide tolerant bio agents:** Biocontrol agents are very susceptible to insecticides and abiotic stresses, though in a totally insecticide-free environment, they have been reported to be effective to the tune of 50-60%. Hence, by developing Multiple insecticide resistant strains. The Multiple insecticide resistant and temperature resistant strains of *Trichogramma chilonis* and multiple insecticides resistant strain of *Chrysoperla zastrowi sillemi* can be used in insecticides and high temperature (ICAR-NBAIR).

vii) Management practices like using organic manures, and parasitoids can be conserved if crop residues are left unburned.

With the increased need and awareness of integrated pest management concepts among farmers, there is increased emphasis on the utilization of biocontrol agents for the management of pests. Though their demand is increasing, their availability is far from sufficient. The biocontrol agents, particularly parasitoids and predators, have short life spans and they cannot be stored for long. Their transportation also requires certain specific conditions which are difficult and expensive. These are some of the important reasons that private sectors are not much interested in them. Of late, there have been quite a few inventions for the automation in the mass production of biocontrol agents which have removed bottlenecks for their economic and efficient production and have facilitated the mass production of biocontrol agents. Now, the biocontrol agents can be mass produced at a small scale/cottage industries just on the lines of sericulture or apiculture. Simple, low cost and down-to-earth technology can be used for their mass production. They can be marketed in the region where they are produced. Simple methods suitable for their transportation have been developed. The production of biological control in cottage industries is also likely to increase their acceptability in rural areas.

References

Arim, M. and Marquet, P. A. 2004. Intraguild predation: A widespread interaction related to species biology. *Ecology Letters* **7**:557–564.

Baker, K. F. and Cook, R. J. 1974. Biological control of plant pathogens. American Phytopathological Society, St. Paul, MN. pp 35-50.

Barratt, B. I. P., Moran, V. C., Bigler, F. and Van Lenteren, J. C. 2018.The status of biological control and recommendations for improving uptake for the future. *Biocontrol.* **63**: 155–167.

Batalden, R.V., Oberhauser, K. and Peterson, A. A. 2007. Ecological niches in sequential generations of eastern North American Monarch butterflies (Lepidoptera: Danaidae): the ecology of migration and likely climate change implications. *Environmental Entomology.* **36**: 1365–1373.

Beddington, J. R., Free, C. A. and Lawton, J. H .1978. Characteristics of successful natural enemies in models of biological control of insect pests, *Nature*: 978 Jun 15; 273 (5663):513-9.

Carriger, J. F., Rand, G. M., Gardinali, P. R., Perry, W. B., Tompkins, M. S. and Fernandez, A. M. 2006. Pesticides of Potential Ecological Concern in Sediment from South Florida Canals: *An Ecological Risk Prioritization for Aquatic Arthropods. Soil and Sediment Contamination: An International Journal,***15**(1):21–45.

Gardiner, M. M. and Landis, D.A. 2007. Impact of intraguild predation by adult *Harmonia axyridis* (Coleoptera: Coccinellidae) on *Aphis glycines* (Hemiptera: Aphididae) biological control in cage studies. *Biological Control* 40: 386–395.

Gill, H. K. and Garg, H. 2014. Pesticides: Environmental Impacts and Management Strategies.In and others (Ed.), Pesticides - Toxic Aspects. *IntechOpen.* https://doi.org/10.5772/57399.

GOI 2021. Insecticides / Pesticides Registered under section 9(3) of the Insecticides Act, 1968 for use in the Country : (As on 01.03.2021)

Gupta, A., Sampathkumar, M., Mohan, M., Shylesha, A.N., Venkatesan, T., Shashank, P. R., Dhanyakumar, O., Ramkumar, P., Sakthivel, N. and Geetha, B. 2021. Assessing adverse impact of the native biological control disruptors in the colonies of the recent invasive pest *Phenacoccus manihoti* Matile-Ferrero (Hemiptera: Pseudococcidae) in India. Global Ecology and Conservation, 32, e01878. https://doi.org/10.1016/j.gecco.2021.e01878

Harper, G. L., King, R. A., Dodd, C. S., Harwood, J. D., Glen, D. M., Bruford, M. W. and Symondson, W. O. C. 2005. Rapid screening of predators for multiple prey DNA targets. *Molecular Ecology* .**14**: 819-828.

Harwood, J. D., Sunderland, K. D., and Symondson, W. O. C. 2004. Prey selection by linyphiid spiders: molecular tracking of the effects of alternative prey on rates of aphid consumption in the field. *Molecular Ecology* **13**: 3549-3560.

Polis, G. A., Myers, C. A. and Holt, R. D. 1989. The ecology and evolution of intraguild predation: Potential competitors that eat each other. *Annual Review of Ecology and Systematics* **20**: 297–330.

Thompson, L., Macfadyen, S. and Hoffmann, A. 2010. Predicting the effects of climate change on natural enemies of agricultural pests. *Biological Control.* 52: 296-306.

Waage, J. and Greathead, D.J. 1988. Biological control: challenges and opportunities. *Philosophical Transactions of the Royal Society B,* **318**: 111-128.

Chapter-4

BIO CONTROL BASED IPM STRATEGIES AGAINST MAJOR INSECT PESTS AND DISEASES OF COCONUT

V. Govardhan Rao, V. Anoosha, B. Neeraja, A. Kireeti, G. Koteswara Rao, and N.B.V. Chalapathi Rao

Dr YSRHU-Horticultural Research Station
Dr. YSR Horticultural University
Ambajipeta, Dr. B. R. Ambedkar Konaseema District, A.P.

Introduction

Coconut (*Cocos nucifera* L.) is the most valuable palm which occupies a unique position among plantation crops. Coconut cultivation is primarily confined to the four southern states and in Andhra Pradesh, coconut is grown in an area of 1.13 lakh hectares producing 16,348 lakh nuts with a productivity of 14,500 nuts per hectare. Of these, erstwhile East Godavari district (Dr B.R. Ambedkar Konaseema district, Kakinada district and East Godavari district) contributes 50 per cent in area (54,539 ha) followed by erstwhile West Godavari (24,975 ha), Srikakulam (15,576 ha) and Visakhapatnam (7761 ha) districts.

Among them the insect pests Rhinoceros beetle, *Oryctes rhinoceros* Linnaeus; Red palm weevil, *Rhynchophorus ferrugineus* Olivier; Black headed caterpillar, *Opisina arenosella* Walker; Coconut slug caterpillar *Macroplectra nararia* Moore and coconut eriophyid mite, *Aceria guerreronis* Keifer are prominent and Important . Recently rugose spiralling whitefly, *Aleurodicus rugioperculatus* Martin entered India during 2016. These insect pests damage coconut during different stages and cause economic loss to varying degrees. Bio-control mainly involves identification of natural enemies, rearing them under laboratory conditions, releasing them in the field and preserving the field released natural enemies. Once the released natural enemies establishes in the field, then it maintains the general equilibrium of the pest population. In case of coconut, the bio-control agents are already being used and widely commercialized and following bio-control agents are being recommended for the management of insect pests of coconut.

Table-1: List of bio-control agents against major insect pests of coconut

Insect pest	Bio-control agent	Category
Rhinoceros beetle	Release of *Oryctes rhinoceros* nudivirus (OrNV) infected beetles @ 10 - 12 Nos. per ha.	NPV (Nuclear Polyhedrosis virus)
	Treating breeding sites with green muscardine fungus, *Metarhizium anisopliae* @ 5x 10^{11} spores per m^3.	Entomopathogenic fungi
Black headed caterpillar	Release of larval parasitoids, *Goniozus nephantidis* @ 20 parasitoid / palm and *Bracon brevicornis* @ 30 parasitoid / palm.	Parasitoids
Eriophyid mite	Talc based formulation of *Hirsutella thompsoni* @ 20 g/l/ palm containing 1.6 x 10^8 cfu with a frequency of three sprayings per year.	Entomopathogenic fungi
Slug caterpillar	*Pediobius imbrues*, *Euplectrus* sp and *Euplectromorpha* sp (Larval); *Eurytoma tatipakensis* (Pupal)	Parasitoids
Rugose spiralling whitefly	*Encarsia guadeloupae*	Parasitoid
	Apertochrysa astur	Predator

Use of the biocontrol agents is easy and feasible in case of leaf eating insect pests like black headed caterpillar and slug caterpillar. In this chapter nature of damage and biological control of major insect pests of coconut are discussed.

Rhinoceros beetle

Rhinoceros beetle occurs throughout the year. But the peak occurrence of the adult beetle visiting the palm crowns is from June to September. Population of the beetle was observed to be high in young plantations with many breeding sites and high rainfall. The affected frond, when fully opened, shows characteristic geometric cuts. Holes are present on the unopened spindle leaves, unopened spathes show round to oblong holes. Usually one or two leaves show the cut marks but under severe conditions all the leaves and tender portion of the crown shows injuries made by the beetle; chewed fibrous material is present in the crown or near the bored hole. The adult beetle damages the unopened leaves and spathes. The beetle causes up to 10 per cent loss in yield. Generally palms aged 3 to 10 years are more prone to infestation by the beetle. Moreover, injured portion may attract pests such as red palm weevil and also pave way for fungal infection (bud rot). Repeated infestation to the growing points may eventually lead to the death of seedlings.

Life Cycle of Rhinoceros Beetle

The female beetle lays eggs after mating in rotting palm trunks or other organic matter (palm debris, compost). The eggs are laid– approximately three per beetle per week, resulting in an estimated 100 eggs per female. The eggs are white, elliptical (3.5 x 4.0 millimetres) and produce the first instar larvae in 8–12 days. Larvae are C-shaped, whitish with a reddish-brown head capsule with sclerotised mouthparts.The larvae grow rapidly, passing through each of the first and second instars in approximately 2—3 weeks, before entering the longer third instar stage (9–24 weeks) where the larvae accumulate most of their mass, growing up to 100 millimetre in length Prior to pupation, the larvae stop feeding and their bodies change to a creamy colour and shrink slightly. The pre-pupal larva forms a cell and differentiates into the pupal stage which lasts for approximately three weeks before metamorphosising into the adult beetle. The adult beetle emerges from the puparium and remains in the feeding site for several days.

Integrated management practices for rhinoceros beetle *Oryctes rhinoceros*

- ☆ **Sanitation:** Remove and destroy all possible decaying debris and dead palms from the plantation since this act as prolific breeding grounds for the beetle.
- ☆ **Mechanical:** The adult beetles may be extracted using curved beetle hooks from the palm crown particularly during the peak period of population build up. The holes made by the beetles are to be filled with a mixture of neem seed kernel powder 100g + 150g sand.
- ☆ **Prophylactic leaf axil filling:** To protect the young palms from rhinoceros beetle attack, the innermost 2-3 leaf axils are to be filled with a mixture of neem seed kernel powder 100g + fine sand (150g) per palm during May, September and December.
- ☆ **Biological management with the entomo pathogen *Metarrhizium anisopliae* (Metch.) Sorokin:** This fungus can be mass cultured in rhinoceros grubs or broken maize grain or coconut water or on cassava chips and rice bran supplemented with a nitrogen source. Periodical spraying of fungus on manure heaps is to be done during monsoon season. The fungus should be inoculated @ 5 x 10^{11} spores/ m^3 of the breeding material.
- ☆ Placing of Napthalene balls @ 12 gm per palm in leaf axils will also deter beetles.
- ☆ **Traps for attraction:** Establishment of rhinolure traps @ 1 trap/ha. will help in effective trapping of adult beetles.
- ☆ Regular monitoring for any damage in the spindle region of the juvenile palm.

Red Palm weevil- *Rhynchophorus ferrugineus*

Symptoms of Damage

- ☆ The hole can be seen on the stem with chewed up fibres protruding out.
- ☆ Many times reddish brown liquid can be seen oozing out from the hole.
- ☆ The grubs cause damage inside the stem or crown by feeding on soft tissues and often cause severe damage especially when a large number of them bore into the soft, growing parts. In case of severe infestation the inside portion of trunk is completely eaten and become full of rotting fibres.
- ☆ In case of young palms the top withers while in older palms the top portion of trunk bends and ultimately breaks at the bend (wilting).
- ☆ Sometimes the gnawing sound produced by the feeding grubs inside will also be audible.
- ☆ In the advanced stage of infestation yellowing of the inner whorl of leaves occur. The crowns falls down or dry up later when palm is dead.

Life cycle of Red palm weevil

After fertilization, the adult female can lay between 300 and 500 eggs. They lay in holes they produced while searching for food. After laying, the female protects and secures the eggs with a secretion that rapidly hardens around the eggs. On average, females produce 210 eggs per clutch on new growth in the crown of the palm, at the base of young leaves, or in open lesions on the plant. The egg hatches into a white, legless larva. The larva will feed on the soft fibres and terminal buds, tunneling through the internal tissue of the tree for about a month. The larvae can occasionally grow to a length of six to seven centimeters. At pupation, the larva will leave the tree and form a cocoon built of dry palm fibers in leaf litter at the base of the tree or in the decaying matter in the palm. The total life cycle takes about 7–10 weeks.

Integrated Management Practices for Red palm weevil

1. The palm crown has to be cleaned periodically to avoid decaying of organic debris in leaf axils. Dead palms which lodge various stages of the weevil should be removed, cut open and burnt so as to destroy all stages of the pest there by preventing spread of the weevil to neighboring healthy palms.
2. As far as possible avoid making any cuts causing injuries to the palm through agricultural tools and implements as these will attract weevil for egg laying.
3. Chisel out affected trunk region and burn it and smear the wounded portion with coaltar.
4. The commercially available pheromone lure is hung inside from the lid of the bucket which is filled with 100 gm pineapple/sugarcane, 2 gm yeast

and in half litre of water along with 0.1 ml Imidacloprid as food bait which is necessary to orient the weevil into the trap. The trap is hung at a height of 1 to 1.5 m above the ground level in the field @1 trap per hectare.

Black Headed Caterpillar- *Opisina arenosella* Walker

It is one of the major factors that contribute to the loss in coconut production and productivity (Ramachandran *et al.,* 1979). The caterpillar lives on the lower surface of leaflets in galleries made of excreta and silken web and feeds on parenchymatous tissues. Continuous feeding removes superficial tissues of the abaxial (lower) leaf surface leaving only the top layer and the leaf lamina which appears white (Howard, 2001). Dried patches appearing on the upper epidermis of leaves and the presence of larval galleries and pupal cases on the lower surface of leaves are the major symptoms of infestation. During severe outbreaks, it feeds even on green surfaces of petioles, spathes and nuts and whole affected plantation presents a burnt up appearance due to drying of leaves. The attacked leaves droop, the bunches buckle and the immature nuts shed heavily (Muthiah, 2007). The other hosts for this pest are palmyrah (*Borassus flabellifer*), talipot (*Corypha umbraculifera*) and wild date (*Phaoenix sylvestris*). Low feeding damage by this pest on oil palm (*Elaeis guineensis*) was also recorded. The cropping system for coconut trees on fish pond bunds prevalent in the coastal ecosystem provides highly favourable climatic conditions for multiplication of black-headed caterpillar and subsequent outbreaks.

Life cycle of black headed caterpillar

The female moth lays eggs in small groups on the underside of coconut leaflets near to feeding larvae. The larvae hatch from the eggs after three days on average. Larvae of *O. arenosella* generally have been recorded to go through as many as eight instars which are completed in about 42 days, including a distinct pre-purpal period. Pupation occurs in the galleries which last approximately eight days. The pupal period is about 12 days. The adult longevity is 5-7 days.Egg to adult stage is completed in 2-2 1/2 months.

Management

From different field experiments and practice, it is proved that, this intricate pest problem can be brought under check by an Integrated Pest Management programme, which is given below.

i. **Mechanical:** Early to mild stages of infestation can be reduced by cutting and burning the badly infested leaves/ leaflets.
ii. **Biological:** Among the parasitoids, the larval parasitoids, *Bracon hebetor, Goniozus nephantidis, the* prepupal parasitoid, *Elasmus nephantidis* Rohw and pupal parasitoid, *Brachymeria nosatoi* are the most promising ones.

These parasitoids are being mass multiplied and released in fixed norms in *Opisina arenosella* affected coconut plantations. Recommended release rates of *G.*

nephantidis and *B. hebetor* were 10 parasitoids per palm and 20 parasitoids per palm, respectively. In addition to above parasitoids other effective parasitoids that are being observed and collected in the infested gardens under east coast conditions are the larval parasitoid, *Apanteles taragamae* and pupal parasitoids, *Xanthopimpla punctata, Trichospilus pupivora, Tetrastichus israeli* etc., the bacterium *Serratia marcescens* kills the pest during the rainy seasons.

Besides these parasitoids the other natural indigenous mortality factors such as carabid predators (*Parena nigrolineata & Calleida splendidula*), anthocoreid (*Cardiastethus* spp.), chrysopid (*Ankylopteryx octopunctata candida*) and several species of spiders (*Cheiracanthium, Heteropoda laprosa* Simon, *Morpisa calcutaenisis* Tikader, *Melanostoma* Thorell, *Olios lamarcki* Latreille, *Rhene* and *Sparassus* spp.) take a heavy toll of different stages of this pest in the field and exert an appreciable effect on pest population.

Under Epidemic populations

- ☆ Removal and burning of one or two badly infested leaves in the outermost whorl.
- ☆ Release of larval parasitoids and pupal parasitoids.
- ☆ Combined with the management techniques, adequate manuring (recommended dose of fertilizers: 1 kg urea, 2 kg single super phosphate and 2 ½ kg muriate of potash/palm) and irrigations are to be resorted for rejuvenating the very severely affected palms.

The periodic out breaks of coconut black headed caterpillar in coastal coconut plantations of Andhra Pradesh state is becoming a cause of concern to the plantation farmers and management of these out breaks with insecticides is having undesirable attributes *viz.,* environmental pollution, residues, resistance, side effects on non-target organisms and the natural enemies of pests. In case of severe out breaks the financial loss suffered by the farmer is also huge and this hampers his investment strength of providing his regular farm inputs to his coconut plantations ultimately affecting the production. The large scale timely availability of bio agents will strengthen and boost the confidence of farmers on biological control. The large scale success of biological control as an alternative system to insecticide usage gives impetus to sustainable agriculture coupled with reducing the cost of cultivation and can be a viable step for doubling the farmer's income if ably supported by governmental organisations, institutions in association with farmer societies.

Coconut slug caterpillar *Macroplectra nararia* Moore (Limacodidae: Lepidoptera)

The out break of this pest on Coconut in East Godavari district was earlier reported by Sujatha *et al.* (2008) and on Oil palm in West Godavari district by Kalidas (2002). Feeding by the early instar caterpillars causes leaf spots and grown up caterpillars eat away entire laminar portion of the leaf leaving the mid ribs. Scorched/ burnt appearance of leaves is the predominant symptom observed in the field during severe infestation. Most of the times, balls of excreta will be seen as a layer on the ground around the coconut palm basin. In severe cases, the pest attacks even nuts and leaf stalks too. During out breaks, all the functional leaves are dried leaving only the spindle leaves, which results in pre-mature drooping of leaves and shedding of nuts, ultimately resulting in yield losses and delayed spathe emergence (Sujatha *et al.*, 2008; Rajan *et al* ., 2011). The pest was also observed to cause damage to the intercrops banana and cocoa in coconut based cropping system in severe out breaks.

In spite of availability of chemical control measures for the management of coconut slug caterpillar the residual toxicity and large scale adoption cost limits its usage in coconut plantations. This situation warrants immediate attention to develop eco friendly and economical method of pest control. Artificial night light attracts many moths, especially light with high ultraviolet (UV) emission (Frank, 1988; 2006). Studies by Sujatha *et al.*(2011) revealed that slug caterpillar is positively phototactic and placing of various light traps at 1½ feet above ground level in coconut slug caterpillar infested villages in East Godavari district of Andhra Pradesh had attracted high number of slug moths . This positive phototactic phenomenon can be exploited for this pest for monitoring, mass trapping and destruction, which is an economical and easy method of pest control.

Life cycle of coconut slug caterpillar

About 200 eggs are laid on the under surface of leaves in batches of 5 to 10 and egg period 6-7 days. The larval Larval period is about 25- 35 days and it has greenish body has rows of spiny scoli which cause irritation and pain when touched. It pupates in a compact elliptical chocolate brown shell like cocoon, and the pupal period 21 days.

Management

- ☆ Monitor the crop in the summer months for pest incidence by installing light traps.
- ☆ Cutting and burning of severely infested and dried leaves will reduce the pest population to a greater extent.
- ☆ Immediate application of recommended dose of fertilizers (1 kg Urea, 2 kg Single super phosphate and 2 ½ kg Muriate of potash) and irrigation will boost the palms t o recover early.
- ☆ Monitoring the pest incidence on other inter crops also helps to keep the pest under check.
- ☆ Installation of light traps [200 w bulb] for mass trapping and destruction of adult moths.

Management of Slug caterpillar with bio-control agents

Paecilomyces lilacinus is a potential pathogen against coconut slug caterpillar (Snehalatharani *et al.*, 2012).

Natural abundance of slug caterpillar parasitoids

Larval parasitoids *Euplectrus* spp. and *Euplectromorpha* spp. belonging to family Eulophidae (Hymenoptera) and a pupal parasitoid *Eurytoma tatipakensis* were effective against slug caterpillar.

Pediobius imbreus as biocontrol agent

Biologically, species of *Pediobius* are quite diverse, acting as primary or secondary parasitoids, utilizing eggs, larvae and/or pupae of species in the insect orders Coleoptera, Diptera, Hemiptera, Hymenoptera, Lepidoptera, Mantodea and Thysanoptera (Boueek 1988; Thangjam *et al.*, 2013). *P. imbreus* is reported to be an important parasitoid on bag worm *Metisa plana* on oil palm while Basri *et al* (1995) reported that *P.anomalus* and *P. imbreus* are the common hyper parastoids on *Dolichogenidea metesae* the most pronounced larval parasitoid of *M. plana*. *P. imbreus* is also reported to be a hyperparasite of *B. brevicornis/ B. hebetor* an important prepupal parasitoid of *O.arenosella*. (Ghosh and Abdurahiman, 1988).

Efficacy of *P. imbrues* on slug caterpillar under laboratory conditions

Laboratory observations on the parasitisation of *P. imbreus* on slug caterpillar revealed that slug caterpillar larvae (0.6 to 0.8cm size) were effectively paralysed with in 48 to 72 hours and egg laying with 3 to 4 eggs in each paralysed larvae was observed on caterpillars of earlier instars (less than 0.6 to 0.8 cm size) and late instars (size above one cm) even though stinging and paralysation was observed no egg laying was recorded. Emergence of parasitoids progeny was observed from 12 to 15 days after parasitisation. Under laboratory conditions longevity of *P. imbrues* parasitoids ranged from 60 to 70 days.

Greater coconut spike moth *Tirathaba rufivena* (Pyralidae, Lepidoptera)

Incidence of greater coconut spike moth *Tirathaba rufivena* is also being observed to be increasing especially in young and dwarf coconut palms in East Godavari district of Andhra Pradesh. The coconut spike moth *T. rufivena* is often referred to as bunch moth in oil palm and is reported as a major insect pest of palmaceae such as *Areca catechu* L. (areca), *Cocos nucifera* L. (coconut), and *Elaeis guineensis* Jacq. (African oil palm) and well distributed in China, Malaysia, Indaonesia, Philippines, and Sri Lanka. (Baozhu Zhong *et al.*, 2017). In India it was reported to cause nearly 6-9 per cent nut loss to the coconut palms in Bastar tribal belt of Chhattisgarh state (Rajesh Kumar Patel, 2018).

In coconut the caterpillar bores at perianth portion in small size to large size buttons and causes nut drop (Nair and Visalakshi, 1999). The affected nuts can be recognized by the presence of excreta and oozing gummy substances. Generally a single larva is found in an infested nut but in case of severe infestation more than one larva can be seen. Male flowers are especially attacked by the larvae which are very active and move quickly when disturbed. Infestation causes abortion of young, underdeveloped nuts. It was reported that the compact leaves of the crown in young (3 to 5-year-old) hybrid palms of the variety MAWA (Malayan Dwarf x West African Tall) in Philippines cause the flowers to be constricted and 100% damage of the female flowers by coconut spike moth was reported in this hybrid (M.J.W. Cock, *et al .,* 1987).

In oil palm on developing bunches, the larvae feed and scrape on the fruitlets, later boring holes into the mesocarp up to the kernel. High infestation can greatly reduce the quality and weight of the fruit bunch and may cause malformed and premature bunch abortion (Alouw *et al.,* 2005). It was also reported that the oil palm plantations which experienced heavy infestation have recorded lower number of bunches per palm and high number of rotten bunches (Idrus *et al.,* 2016). In oil palm infestation is characterised by the presence of long tube of silk and frass in the bunch, which are reddish when fresh and brownish-black when old (Lim, 2012). The oil palm plantations in Malaysia mainly rely on insecticides, particularly cypermethrin, to control the bunch moth (Lim, 2012) which is adversely affecting the population of oil palm pollinating weevil, *Elaeidobius kamerunicus* and its rate of degradation is also slow when applied on peat (Ismail *et al.,* 2012). Efforts to manage this pest using biological agents such as the parasitoid, *Argyrophylax basifulva* and nematode, *Steinernema feltiae* have been unsuccessful (Godfray, 1985; Zelazny, 1985). Hence the present studies on successful identification of a lab amenable parasitoid which can be mass multiplied bear's importance.

Biological control of *T. rufivena*

Laboratory and field studies conducted in the Philippines to assess the effectiveness of different parasites and pathogens against *T. rufivena* revealed that larvae of >15 mm were highly susceptible to the entomoparasitic nematode, *Steinernema feltiae,* in the laboratory (Zelazny, 1985) however; the pest was not significantly reduced in field trials. The studies on the ichneumonid *Venturia palmaris* against *T. rufivena* in Malaysia revealed that the parasite was present throughout the year but no parasitism was observed in November or December months. The degree of parasitism ranged from 2% in August to 31.9% in February. The parasite preferred to oviposit in late-instar larvae rather than in young larvae or pupae. Laboratory studies on life cycle and oviposition behaviour indicated that the sex ratio was probably the limiting factor in mass rearing *V. palmaris* in the laboratory for use in biological control (https://www.plantwise.org/KnowledgeBank/Datasheet.aspx).

The laboratory observations on parasitisation efficiency of *G. nephantidis* on *T.rufivena* if any was conducted from November, 2017 to June, 2018 The parasitoid *G. nephantidis* was reared on the alternate host, *C. cephalonica* and were released in to the test tubes (6"×1") containing *T. rufivena* larva of varying sizes. The results revealed that spike moth larvae (10 to 15 mm size) were effectively paralyzed within in 48 to 72 hours and around 10 to 15 eggs in each paralyzed larvae were recorded . The life cycle of the parasitoid was completed in 10-15 days (incubation period 24 hrs, larval feeding 48-72 hrs, prepupal stage 48 - 60 hrs and cocoon period 48 to 56 hrs + resting adult inside the cocoon 120 hrs) and the progeny obtained were in 1: 3 male female ratio and laboratory longevity of *G. nephantidis* emerging from this pest ranged from 60 to 65 days.

The parasitoid *G. nephantidis* is apparently reported to be host-specific, but may have a broader host range in coconut ecosystem and present study confirms *T. rufivena* as one more alternate host . As this parasitoid is amenable for mass production therefore field evaluation of *G. nephantidis* for management of *T. rufivena* in coconut and oil palm can be contemplated in near future or successful management of this pest.

Rugose spiralling whitefly, *Aleurodicus rugioperculatus* Martin (Hemiptera: Sternorrhyncha: Aleyrodidae)

The increasing demand to gain more production by adopting new technologies and changing agricultural practices results in new hazards to health that are readily able to cross borders. This free trade and movement of plant materials across political and geographical borders has led to the introduction of certain plant, insect and pathogen species to new localities. Such non native species are termed as Alien species. Alien species are non-native or exotic organisms that occur outside their natural adapted habitat and dispersal potential. Many alien species support our farming and forestry systems in a big way. However, some of the alien species become invasive when they are introduced deliberately or unintentionally outside their natural habitats into new areas where they express the capability to establish, invade and outcompete native species. (Raghubanshi *et al.*, 2005). These species, if not accompanied by the natural enemies which keep them in check in their native range, can multiply in large proportion and cause damage to economically important plant species and crop plants which makes them invasive. Some examples are invasion of coconut eriophyid mite *Aceria guerreronis*, the cotton mealy bug *Phenacoccus solenopsis*, the papaya mealy bug *Paracoccus marginatus*, the eucalyptus gall wasp *Leptocybe invasa* and rugose spiralling whitefly *Aleurodicus rugioperculatus* Martin (Hemiptera: Sternorrhyncha: Aleyrodidae) (Ananthakrishnan, 2009). Recently, fall army worm, *Spodoptera frugiperda* (J.E. Smith) has been reported from maize fields in Karnataka (Ganiger *et al.*, 2018). Furthermore, ICAR-CPCRI has discovered a new invasive Bondar's Nesting Whitefly (*Paraleyrodes bondari* Peracchi Hemiptera: Aleyrodidae) from Kerala during December 2018. It was recorded from coconut gardens earlier infested by rugose spiralling whitefly (*Aleurodicus rugioperculatus* Martin) during 2016 (Anonymous, 2019).

International Union for Conservation of Nature and Natural Resources (IUCN) defines Alien Invasive Species as 'one which becomes established in natural or semi natural ecosystems or habitat, and threatens native biological diversity'. These invasive species are widely distributed in all kinds of ecosystems throughout the world and include all categories of living organisms. Nevertheless, plants, mammals and insects comprise the most common types of invasive alien in terrestrial environments. Many non-native plants have allowed increasing productivity in agriculture, becoming a fundamental part of human economy, whereas others have developed into serious ecological problems. However, besides non-native plants, potential bio-safety risk factors comprise two additional groups: (a) taxa resulting from traditional breeding, and (b) genetically modified plants (GMPs) (Hoenicka and Fladung, 2006). The spread of Invasive Alien Species (IAS) is now recognized as one of the greatest threats to the ecological and economic well being of the country.

These species are causing enormous damage to biodiversity and the valuable natural agricultural systems upon which we depend. The direct and indirect health effects are increasingly becoming serious and the damage to nature and environment is often irreversible. The effects are exacerbated by global change and chemical and physical disturbance to species and ecosystems.

The Rugose Spiraling Whitefly, *Aleurodicus rugioperculatus* Martin (Hemiptera: Sternorrhyncha: Aleyrodidae) entered India infesting coconut plantations across Kerala, Tamil Nadu, Karnataka and Andhra Pradesh. Presently, infestation of RSW increased over the time and spread across the southern states in India and extending its host ranges at greater level which could be due to its polyphagous nature. Immature stages of RSW produce profuse quantity of wax filaments both tufts of fluffy and long crystal like glassy rods. Furthermore, RSW produce honey dew which results in development of sooty mould. The severity of infestation ranged between 40-60 % in coconut and 25-40 % of leaf in banana and population including all the immature stages varied from 18 to 43 nymphs/sq.cm of leaf and very high at midrib region. Complete drying of banana leaves was also noted in several places in Tamil Nadu and Kerala (Selvaraj *et al.,* 2016).

Origin and Distribution

Rugose spiralling whitefly was initially reported from Miami-Dade County, Florida, United States of America from gumbo limbo, *Burera simaruba* (L.) Sarg in 2009 as a pest. However, it was originally described from Belize in 2004 on coconut (Martin, 2004) where its natural population was reported. This whitefly is believed to have originated from Central America and distribution of this pest in Central and North America is limited to Belize, Mexico, Guatemala and the United States (Evans 2008). In the continental United States, the first established population of rugose spiraling whitefly was reported from Florida in 2009, and since then its distribution range has expanded considerably within the state and subsequently, it has spread to 22 other countries in Central and South America, including Florida, USA. India is the only country in the Oriental region where the whitefly has been introduced. Initially, this whitefly was observed in several coconut farms in the Pollachi area of Coimbatore district, Tamil Nadu and first reported in Kottayam from Kerala during July – August 2016 (Sundararaj and Selvaraj, 2017). RSW was noticed on coconut palms, mango and guava at Changanassery, Kottayam District, Kerala in India through accidental introduction. The pest has also been recorded from in Kadiyapulanka nurseries in Coastal Andhra Pradesh during October-November, 2016. The possible entry to Andhra Pradesh may be by coconut seedlings from nurseries in Tamil Nadu and Kerala.

Description and Biology

Rugose spiralling whitefly was first described by Martin in 2004 from samples collected in Belize on coconut palm leaves (Martin, 2004).

Adults: Rugose spiralling whitefly adults are about three times larger (approx. 2.5 mm) than the commonly found whiteflies and are lethargic by nature. Rugose spiraling whitefly adults can be distinguished by their large size and the presence of a pair of irregular light brown bands across the wings (Stocks and Hodges, 2012). Males have long pincer-like structures at the end of their abdomen.

Eggs: Females lay eggs on the underside of leaves in a concentric circular or spiral pattern and cover it with white waxy matter. Eggs are elliptical and creamy white to dark yellow in color.

Immature stages: Rugose spiralling whitefly has 5 developmental stages. The first instar, known as the crawler stage (because it is the only mobile immature stage) hatches out of the egg, and looks for a place to begin feeding with its needle-like mouth parts and sucks plant sap. Crawlers molt into immature stages that are immobile, oval and flat initially but become more convex with the progression of its life cycle (Mannion, 2010). Nymphs are about 1.1 - 1.5 mm long but may vary in size depending on instars. The nymphs are light to golden yellow in color, and will produce a dense, cottony wax as well as long, thin waxy filaments (Stocks and Hodges, 2012) which get denser over time.

Puparium: This species is typically characterized by broadly cordate vasiform orifice, operculum ventro-basally spinulose and dorsally characteristically rugose, with a pair of ventro-median fine setae; lingula head protruding beyond vasiform orifice, finely spinulose, apically acute, its four setae situated close to apex.

Host Range

It is a polyphagous pest feeding on a wide range of host plants including palms, woody ornamentals, and fruits (Mannion 2010). Florida Department of Agriculture and Consumer Services (FDACS), Division of Plant Industry (DPI) records from 2009 to 2015 identified rugose spiraling whitefly on at least 118 plant species, which include a combination of edibles, ornamentals, palms, weeds as well as native and invasive plant species (Stocks, 2012). Further, host plants recorded from 2009 to 2012 at Florida shows that 22% of rugose spiraling whitefly affected hosts were palm species, 16% were gumbo limbo, 10% were *Calophyllum* spp., 9% were avocado, 4% were black olive and 3% were mango varieties (Francis *et al.*, 2016). Within the family Arecaceae (palms), 44% of host records were from coconut. Based on incidence records, these plant species can be considered as primary or preferred hosts of this pest. However, all plant species reported have not been documented as true hosts of the pest and may not require management. An insect must be able to complete its entire life cycle (egg to adult) to be considered a true host plant. Some plant species may not support the complete development of rugose spiraling whitefly but may still be used by adult whiteflies for feeding and laying eggs. Thus, the level of feeding by adult whiteflies and development of other stages will determine the impact the whitefly has on the host plant and if management is required. A total of 17 plant species under 11 families were recorded as preferred hosts of *A. rugioperculatus* at Kerala (Shanas *et al.*, 2016). The host preference of A. *rugioperculatus* was also recorded by scoring the presence of live egg spirals on leaf and categorized as Low «10 egg spirals/ leaflet), medium (10-20 egg spirals/ leaflet) and high intensity (>20 egg spirals/ leaflet).

Table 10: Incidence and intensity of *A. rugioperculatus* on various plants in Andhra Pradesh

S. No	Common name	Scientific name	Spirals per leaflet/ leaf	Intensity
1	Coconut	*Cocos nucifera*	>30	High
2	Oil palm	*Elaeis guineensis*	>30	High
3	Cocoa	*Theobroma cocoa*	<10	Low
4	Banana	*Musa* sp	10 -20	Medium
5	Seethaphal	*Annona squamosa*	Spirals on entire leaf	High
6	Curry leaf	*Murraya koenigii*	Spirals on entire leaf	Medium
7	Jack fruit	*Artocarpus heterophyllus*	Lower no of spirals	Low
8.	Papaya	*Carica papaya*	<10	Low
9	Yam	*Colacasia* sp	<10	Low
10	Mango	*Mangifera indica*	Lower no of spirals<10	Low
Ornamentals				
11	Bird of paradise	(*Strelitzia reginae*)	<10 (Low)	
12	Fish tail palm	*Wodyetia bifurcata*		
13	Spider lily	*Lycoris* sp		
14	Areca palm	*Chrysalidocarpus lutescens*		
15	Cabbage tree	*Pisonia alba*		
16	Rose apple	*Syzigium malaccense*		
17	Heliconia	*Heliconia stricta var. Iris Red*		

Biological control of RSW

Parasitoids *viz., Encarsia guadeloupae* Viggiani (Hymenoptera: Aphelinidae) was known to parasitise *A. rugioperculatus* while Poorani and Thanigairaj, 2017 reported *Encarsia dispersa* Polaszek parasitizing *A. rugioperculatus* in surveys conducted at Tamil Nadu. A heavy parasitisation ranging from 40 to 70% was recorded on banana alone by *E. guadeloupae* (Poorani and Thanigairaj, 2017). However, survey conducted by Selvaraj *et al.* (2016) recorded 20–60% parasitism of *A. rugioperculatus* by *E. guadeloupae* on coconut in Tamil Nadu and Kerala. Among the two parasitoids, *E. guadeloupae* was more predominant, causing 60–70% overall parasitism while *Encarsia dispersa* was found in much fewer numbers compared to *E. guadeloupae* and the extent of parasitism was <5% (Poorani and Thanigairaj, 2017). A brief history of introduction and nomenclature and diagnostic details of *E. dispersa* are given in this chapter with notes on the predators of the rugose whitefly.

Encarsia guadeloupae a Potential Parasitoid on Rugose Spiraling Whitefly

Encarsia guadeloupae was first reported from Minicoy Island in the Lakshadweep in 1999 and later deliberately introduced into the mainland and established there (Ramani, 2000; Mani *et al.*, 2000a, b; Beevi and Lyla, 2001).

The parasitoid *Encarsia guadeloupae* was reported as a promising bio agent against RSW. This parasitoid migrated from the Maldives into Minicoy and other islands of the Lakshadweep in 1999 and and later, assisted by the intentional release and colonization, spread to other areas of peninsular India. It is also reported that the parasitoids were found only after their numbers increased phenomenally through breeding for several years on the expanding host population, although they had been introduced along with the host *A. dispersus.*

Key diagnostic characters of female *E. guadeloupae* are: body dark brown, except side lobes of mesoscutum and scutellum; mid lobe largely dark brown; axillae brown; TVII yellow or brown laterally; third valvula pale yellow. Wings hyaline, hind coxa and femur dark brown, legs otherwise pale yellow to white. Antennal formula, 1142; clava as long as last two funicle segments; mesoscutum with usually 9 pairs of setae (sometimes with variable number of setae, 15–20); scutellum with two pairs of setae; axillae normally with one setae each.

Fore wing basal cell with 3 setae before parastigma. Mid tarsi-4 segmented; fore and hind tarsi-5 segmented. Similarly, *E. guadeloupae* and *Encarsia* sp. were recorded as potential parasitoids of many whiteflies including RSW (Evans, 2008: Taravati *et al.*, 2013; Francis *et al.*, 2016). Furthermore, the COI gene (658 bp) of the parasitoid was amplified, sequenced and deposited as *Encarsia* sp. (GenBank Acc. No. KY223606). The per cent parasitism ranged from 20.0-60.0 % in different locations, with highest parasitism recorded in Kerala as compared to other surveyed states (Selvaraj *et al.*, 2016).

Encarsia Dispersa (Polaszek)

Encarsia dispersa Polaszek, is an exotic parasitoid of New World (Neotropical) origin. This species has been widely and erroneously referred to as *Encarsia haitiensis* Dozier, *Encarsia* sp. nr. *haitiensis* and occasionally as *Encarsia* sp. nr. *meritoria* Gahan by different authors from India (Ramani *et al.*, 2002) and elsewhere. It has been both deliberately and fortuitously introduced around the world for the biological control of spiralling whitefly, *Aleurodicus dispersus* Russell (Polaszek *et al.*, 2004). Polaszek *et al.* (1992) regarded *Encarsia haitiensis, Encarsia* sp. nr. *haitiensis* and *Encarsia* sp. nr. *meritoria* as distinct, though very closely related. Based on variations in DNA sequence data for the D2 region of 28S nuclear ribosomal genes, Polaszek *et al.* (2004) formally described this species as *Encarsia dispersa* and included it as part of the *luteola* species-group and the *Encarsia meritoria*-species complex and defined the species limits of *E. haitiensis, E. meritoria* and other related species. The widely introduced populations of *E. dispersa* probably originated from a single, or a few, original populations brought from Trinidad to Hawaii (Polaszek *et al.*, 2004). *Encarsia dispersa* was the first of two species of *Encarsia* accidentally introduced in South India along with its host, *A. dispersus,* in the late 1990s, the other one being *E. guadeloupae* (Ramani *et al.*, 2002). In an apparent case of competitive displacement, *E. guadeloupae* completely displaced *E. dispersa* within a short span of 2–3 years in most of the places in South India where it was colonized (Ramani *et al.*, 2002; Mani *et al.*, 2004; Mani, 2010).

Introduction and Conservation of *E. guadeloupae* in Andhra Pradesh

In Kerala and Tamil Nadu parasitisation range of *E. guadeloupae* was reported from 40 to 70% in banana and 20-60% in coconut on RSW. Intensive surveys carried out by scientists of Dr. YSR Horticultural University in the year 2017 to find the *E. guadeloupae* parasitisation in RSW infested Coconut, Oil palm plantations and nurseries of Kadiyam area in Andhra Pradesh revealed that *E. guadeloupae* parasitisation on RSW was not observed. Hence, with an objective of establishing *E. guadeloupae* the scientists of Dr. YSRHU - HRS, Ambajipeta brought *E. guadeloupae* parasitised white fly pupal stages from TNAU (Tamil Nadu Agricultural University), Aliyarnagar, Tamil Nadu and ICAR-CPCRI, Kasargod, Kerala and ICAR- NBAIR, Bengaluru, Karnataka in a phased manner. The first consignment of 1500 numbers of parasitised white fly pupal stages were released in coconut gardens in Kalavalapalli village of West Godavari district on 18.12.2017 and second consignment of 2500 numbers were released in Oil palm and Coconut gardens on 08.01.2018. These attempts resulted in approximately 72 % field level parasitisation of RSW by *E. guadeloupae*. However *Encarisa* establishment though satisfactory upto March was affected in subsequent months due to higher temperatures In 2018, September *E. guadeloupae* parasitised white fly pupal stages were obtained and re- released in RSW infested coconut gardens and these attempts resulted again in parasitisation of RSW by *Encarsia* (62 per cent). Consignments of *E. guadeloupae* were obtained from TNAU on 2.11.2019 (1500 number parasitised white fly pupal stages) and from ICAR-NBAIR Bangalore on 12.12.2019 (1400 number parasitised white fly pupal stages) and 24.12.2019 (1600 number parasitised white fly pupal stages) and released in identified Goutami Ganga (dwarf variety) block in the research farm (Table-11).

Table.11: Source and Number of *Encarsia guadeloupae* parasitised RSW pupae obtained and released in Andhra Pradesh from 2017 to 2019

Year	Source	Number of *E. guadeloupae* parasitised RSW pupae obtained
December, 2017	TNAU, Aliyarnagar, Tamil Nadu and ICAR-CPCRI, Kasargod, Kerala	1500*
January, 2018	TNAU, Aliyarnagar, Tamil Nadu and ICAR-CPCRI, Kasargod, Kerala	2500*
November, 2019	TNAU, Aliyarnagar Tamil Nadu	1500**
December, 2019	ICAR- NBAIR, Bengaluru, Karnataka	1400**
December, 2019	ICAR- NBAIR, Bengaluru, Karnataka	1600**

*- Released at Kalavalapalli village of West Godavari district.

** -Released at Dr.YSRHU-HRS,Ambajipeta of Dr.B.R.Ambedkar Konaseema district.

The parasitoid population increased and about 48% percent parasitisation was observed even after summer 2020 and which increased subsequently to 88% in December during which period RSW population is favorably building up. In 2021, *E. guadeloupae* acclimatized and established to the local environmental conditions (Parasitisation range 28.92-83.19%) with the peak parasitisation observed in the month of January. Further, in the summer months the parasitoid population did not dwindle and parasitisation percentage of 59.71 % and 34.92% observed during the months of April and May, 2021.

Starting from the January 2021, 15000 *E. guadeloupae* parasitised RSW pupae leaf clippings (Each clipping containing 3-5 parasitised pupae) were re-distributed to the coconut plantation growers from East Godavari, West Godavari, Vishakapatnam and Srikakulam districts for the successful establishment of this parasitoid in those regions. Subsequent surveys in 2022 in these districts revealed that successful *E.guadeloupae* colonization in RSW infested gardens in the redistributed areas.

In 2022, *E. guadeloupae* parasitisation in the range of 19.14 - 48.25% was observed at Dr. YSRHU-HRS, Ambajipeta farm inspite of re-distribution of parasitised RSW pupae leaf clippings in 2021 indicating that once established *E. guadeloupae* can successfully sustain itself in the released plantations. As parasitism levels of *E. guadeloupae* were found to be highly density-dependent the buildup of RSW will increase parasitisation by *E. guadeloupae* proportionately (Table-12).

Table.12: Establishment of parasitoid *E. guadeloupae* in Kalavalapalli village and Dr.YSRHU-HRS., Ambajipeta from 2018 to 2022.

Month	Percent rugose spiraling whitefly pupae parasitised by *E. guadeloupae*				
	2018	**2019**	**2020**	**2021**	**2022**
January	20.01	71.35	4.22	83.19	27.44
February	72.06	65.44	1.49	80.31	34.65
March	52.81	28.68	9.88	74.94	37.25
April	Nil	11.41	18.50	59.71	24.35
May	Nil	8.17	27.81	34.92	20.21
June	Nil	2.53	36.36	31.04	19.14
July	Nil	Nil	47.60	28.95	26.54
August	Nil	Nil	72.71	43.72	32.54
September	29.34	3.39	78.47	39.78	34.75
October	42.38	4.91	66.80	35.11	38.65
November	69.49	7.70	79.71	32.04	42.58
December	68.83	9.75	88.08	28.92	48.25

(Data collected from 10 palms at random) *- 16 leaflets were sampled per palm

Measures to Conserve Parasitoid *E. guadeloupae*

Conservational biological control can be promoted through non usage of pesticides. Hence creating awareness in farmers about deleterious impact of pesticide usage is vital. Among the various intercrops grown in coconut broad leaved plants like Colacasia and Banana which are also alternate hosts of RSW are supporting the development of *E. guadeloupae*. Further it was earlier reported that ornamental plant *Canna indica* served as reservoir plant enhancing the multiplication of parasitoid *E. guadeloupae*. Therefore promotion of intercropping with Colacasia, Banana and border planting with *C. indica* will conserve this potential parasitoid and successfully managing RSW through conservation biological control. Further studies on off season survival of *E. guadeloupae* needs to be explored especially under low RSW population levels. Redistribution of *E. guadeloupae* parasitoid infested RSW pupae leaflets will help in faster spread of the parasitoid and can regulate RSW successfully in newly spread areas.

Predators of *Aleurodicus rugioperculatus*

Many indigenous predators have been observed feeding on *A. rugioperculatus* as in the case of *A. dispersus*. Poorani and Thanigairaj, 2017 collected predators of rugose whitefly on coconut at Kerala and Coconut Research Station (TNAU), Aliyar Nagar, were also brought to our attention. Among these, *Apertochrysa spp*. (Neuroptera: Chrysopidae), *Cybocephalus* sp. (Coleoptera: Cybocephalidae), *Diadiplosis* spp. (Diptera: Cecidomyiidae) and *Jauravia pallidula* Motschulsky (Coleoptera: Coccinellidae) were the most predominant.

Apertochrysa astur

Most lacewings (also) feed on plant material, such as pollen. Particularly green lacewings (Chrysopidae) such as *Chrysoperla carnea, Mallada boninensis,* and *Mallada signata* are mainly predatory and very effective. They are relatively small (length up to 1.5 cm, including wings). Adults are active during the night, and may be attracted by flashlight. Most pesticides and high-copper fungicides are toxic to lacewings. A female lacewing may plant over 100 eggs on plants infested with aphids. These eggs hang on a thin 1 cm long stalk, most often under a leaf. The eggs hatch within 4 days. The larval period lasts 12 days and the pupal period lasts about 9 days. The larvae have spines on their back that they impale the remains of their prey on for camouflage.

The natural population of *A. astur* was observed feeding on the eggs and nymphs of RSWF at Kadiyam and Kadiyapulanka villages of East Godavari, Andhra Pradesh. *A. astur* grubs were collected from nursery at Kadiyapulanka. These grubs were reared at parasite breeding station, Horticultural Research Station, Ambajipeta using corcyra eggs. Grubs were reared in individual vials until pupation and then adults were transferred into the adult rearing cages. Adults insects were reared on artificial protein rich diet which is provided in semisolid paste. This diet consisted of equal parts of yeast, fructose, honey, Proteinex R and water. The adults lay eggs on the brown sheet provided on the lid of the rearing cage. The adults were collected on daily basis and transfered into fresh rearing cage with fresh food. From the old cages, the brown paper sheets along with *A. astur* eggs were removed and made into small pieces with 20-25 eggs. These small cuttings were pinned on the underside of RSWF infested coconut leaves. The eggs of *A. astur* are available in the HRS, Ambajipeta and the price is Rs. 150 for 1000 eggs.

Isaria fumasorosea (Wize)

It is an entomopathogenic fungus, formerly known as *Paecilomyces fumosoroseus*. It is a promising as a biological pesticide with an extensive host range.

Mode of action: When the spores of this fungus come in contact with the cuticle (skin) of the pest such as mites and insect, they germinate and grow directly through the spiracle in the cuticle in to the inner body of their host. The fungus proliferates throughout the insect's body, draining the mite and insect of the nutrients and the infected mite and insect eventually dies. *I.fumasorosea* infects the mites and insects upon contact and does not need to be consumed by the host to cause infection.

Host Range of *Isaria fumosorosea* (Wize): It is commonly found in the soil (Cantone and Vandenberg, 1998) but has been reported on plants, in water, and less commonly, in air on every continent except Antarctica (Zimmermann 2008). It has been isolated from over 40 species of arthropods, representing 10 orders. Some of the more commonly known susceptible organisms include weevils, ground beetles, plant beetles, aphids, whiteflies, psyllids, wasps, termites, thrips, and a wide variety of butterflies and moths (Smith, 1993 and Hoy *et al.*, 2010).

Field experiments were conducted in association with NBAIR- ICAR Bangalore in Madavaraidupalem in East Godavari and Kalavalapalli village in West Godavari district, Andhra Pradesh (two gardens) during November - March, 2018-19 to field evaluate the efficacy of *I. fumosorosea* against rugose spiralling whitefly on coconut (Godavari Ganga hybrid). The results revealed that *I. fumosorosea* Pfu-5 reduced the egg hatching (62-78%), caused mortality on early nymphal instars (52-68%) and late nymphal instars (48-63%). However, during high temperatures (at a temperatures of >35°C) there is a comparatively lower suppression of eggs (18.6%), nymphs and pupae (20.3%).

Good results with I. *fumosorosea* Pfu-5 can be obtained if sprayings are carried out at 10 to 15 days interval with high jet sprayers when there are no high temperatures. Further spraying operations should be initiated very early in the season as and when RSW population is observed as it can have more impact and can reduce and arrest the RSW population build up.

Management strategies recommended for RSW

1. Avoid transportation of coconut seedling or any other ornamental plants from pest infested areas to new un infested area.
2. Release of *Apertochrysa astur* @ 100 to 150 eggs / palm during low incidence and up to 300 eggs/ palm during medium incidence for at least 10 per cent of infested palms particularly in those plantation where spraying is not feasible.
3. Re-distribution of *E. guadeloupae* in RSW infested areas taking due care.
4. Avoiding Pesticides spray.
5. Spraying Azadirachtin @ 1% @ 1 ml per litre along with detergent powder @ 10 gms two / three sprays at 15 -20 days interval.
6. Foliar application of entomopathogenic fungi I. fumosorosea @ $1x10^8$ spores/ml (5 gm /litre along with sticker 2 ml/litre).
7. Intermittent Jet water spray at fortnightly intervals.
8. Installation of yellow sticky traps on palm trunk to attract adult white flies and regular smearing with castor oil at 7 to 10 days interval.
9. Community based approach for management.

Conclusion

Coconut is multipurpose palm and its products can be used as food, coir, wood *etc*. The coconut is widely grown in coastal regions of Andhra Pradesh. The major constraints in the production of coconut are insect pests which cause heavy damage to the palm. Prominent insect pests are Rhinoceros beetle, Red palm weevil, black headed caterpillar, slug caterpillar. Insecticidal sprays are difficult to adopt in coconut because of its height and intercrops. Biological control is the effective method to implement in the coconut ecosystem as it offers long term benefits. There are several instances where the insect pest of coconut were managed by bio-control. Sujatha *et al.*, 2004 recorded natural occurrence of *Metarhizium* and *Baculovirus* in East Godavari, West Godavari and Srikakulam. Chalapathirao *et al.*, 2018 released baculovirus infested rhinoceros beetle in Borivanka, Kuttuma, Bhiripuram, Nagullanka and Atreyapuram villages. The release of baculovirus infested beetles showed a significant reduction in leaf and spindle damage at all five demonstration sites. Further, a steady decrease in leaf damage with progression of time which correlated with release of virus beetles was observed.

Black headed caterpillar was managed biologically through release of *Goniozus nephantidis* and *Bracon hebetor*. A significant decrease in larval population was observed after release of parasitioids in all demonstration villages. However, these releases also enhanced the natural parasitisation. The natural parasitisation percentage was 8.30 to 26.92 % which increase to 18.68 and 34.61 % during 3rd and 6th month after release, respectively.

Slug caterpillar is sporadic pest on coconut and its heavy infestation in recent times raised concern. Chalapathirao *et al.*, 2012, recorded natural occurrence *Paecilomyces lilacinus* on coconut slug and they also confirmed the pathogencity *of P.lilacinus* on slug caterpillar. Further, Chalapathirao *et al.*, 2012 recorded some parasitoids viz., *Euplecturs sp, Euplectro morpha, Eurytoma tatipakensis* and *Pediobius imbrues* from coconut slug caterpillar. Among these *Pediobius imbrues* was reared in the lab using slug caterpillar larvae and *Bracon* pupae. The efficacy of the *P. imbrues* was high under lab conditions. However, field efficacy of the *P. imbrues* was also high with 41.41% parasitisation.

Coconut palm provides niche for many pollinators and parasitoids. There are so many insect species in coconut ecosystem which are not documented and many of them might be potential parasitiods. Effective biological control of the *Opisina* by *Goniozus* and *Bracon* was possible because of community based approach without insecticidal sprays. The initial costs were high but the parasitoid recovery per cent was high after repeated releases. Further, the parasitiod release is easier than the conventional insecticidal sprays in tall palms. A successful case of biocontrol mentioned in this chapter gives hope for further identification and standardization of new parasitoid species in coconut ecosystem.

Bio-control Based Management of Coconut Diseases

Introduction

One of the major factors that contribute to the loss of production and productivity in coconut is the damage due to diseases. Basal Stem Rot (Ganoderma wilt) disease caused by *Ganoderma applanatum* (Pers.) Pat., and *G.lucidum* (Leys) Kant. is the most destructive disease of coconut, as the ultimate infection leads to death of palms. The other important diseases are stem bleeding caused by *Thielaviopsis paradoxa* and bud rot caused by *Phytophthora palmivora* can cause severe yield losses. All the three diseases are caused by fungal pathogens which are soil borne in nature. Soil borne pathogens have a broad host range and persist for longer periods in soil by resistant resting structures.

Disease status in the Andhra Pradesh

As per the data of the survey and surveillance programmes conducted during the past two decades it was recorded the disease incidence of Basal stem rot in Andhra Pradesh ranged from 25.65 to 8.28 percent with a mean of 14.11 percent .The stem bleeding incidence ranged from 10.21 to 4.35 percent with a mean of 4.94 percent. The bud rot disease was mostly observed in the young gardens and nursery which showed from 25.63 to 0.86 percent with a mean of 8.64 percent in the state. The leaf spot diseases observed in traces only.

Disease trend in last two decades

The destructive soil borne disease Basal stem rot had shown the down trend in the last two decades and gradual reduction from 20.60% to 8.28 %. The other diseases stem bleeding and Bud rot also observed in the downtrend from 10.21% - 4.35 %and 25.63% - 0.86% respectively during the last ten years [Fig.1]

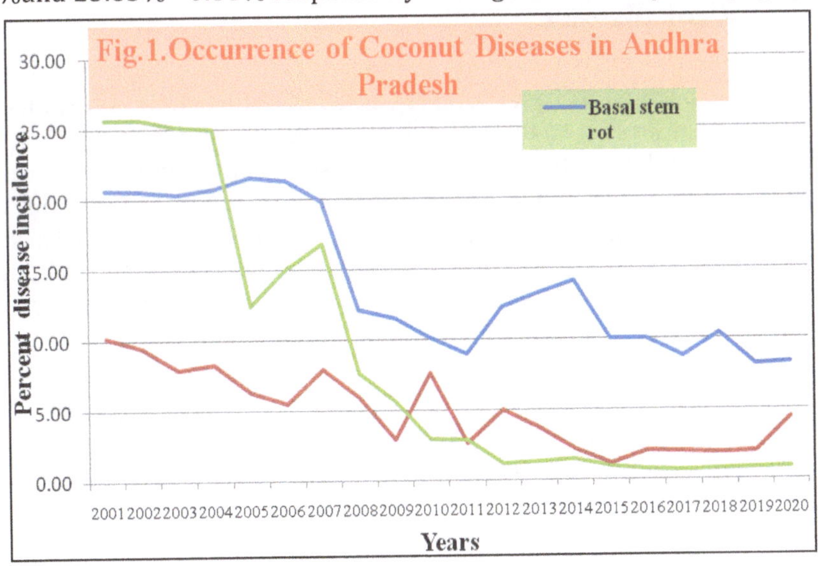

Fig.2.Per cent disease incidence for the last two decades (2001-2020)

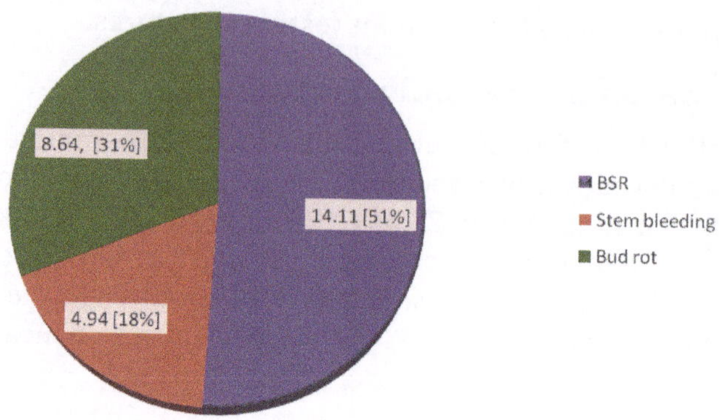

Out of the three diseases 51 % of damage is caused by Basal stem rot damage followed by stem bleeding and bud rot diseases with 31% and 18% respectively in the state [fig.2].

Basal stem rot disease

The Coconut crop grown in Andhra Pradesh is being affected with Ganoderma wilt which is also called as basal stem rot and Tanjavur wilt .It is responsible for considerable yield loss up to 80% and death of palms. About 30% of the annual yield loss is recorded from the sandy soils of coastal belts of Andhra Pradesh where coconut is prominently growing and also decreased crop stand due to this disease. Among the various fungal diseases affecting coconut palm in India, Basal stem rot (BSR) is the most destructive disease. It is the major constraint in lighter soils than in black soils. During recent years, the disease is also found in heavy soils such as black cottony soils and on paddy field bunds.

Symptomology

The disease symptoms expressed on roots, stem and on crown region of the palm. Disease symptoms progress slowly, but usually every infected plant eventually dies. The disease was found to be more in lighter soils than in heavy black soils .However, time period for initiation of disease symptom expression from the infection of palms is varies in black cotton soils and light soils. Disease develops from the roots and the first visual symptoms are visible on stem as reddish brown exudation. Palms aged 10 years and older were more susceptible to the disease than younger palms.

Roots: The pathogen first infects the root system through the wounds and no external disease symptoms are clearly visible during the very early stage of infection. Initially a few roots get infected and rot. Extensive rotting and discoloration of root system is a characteristic symptom of the disease and the rotting proceeds towards the bole thus, cortical tissues disintegrate and the stele turns brown. The production of new roots decreases with increase of disease in the infected palm. In severely infected palms, more than 70 % root rotting was observed.

Stem: From the roots, the infection slowly progresses up the stem leading to internal disintegration of cortical and stele tissues. Exudation of reddish brown viscous fluid from the basal portions of the stem is the first visible symptom of the disease in the affected palm. The disease caused 15 to 25% damage to roots and bole below the ground level by the time external symptoms are visible. The internal tissues of the affected stem turn brown in color and rotting in the stem can be seen up to the height of the bleeding. Bleeding on the stem begins at the base and may extend up to 15 feet in severe cases. Occasionally, some infected palms do not show bleeding symptoms .The bark from the base of the stem peels off. Infestation of scolytid beetle, *Xyloborus perforans* and the weevil, *Diocalandra stigmaticollis* are found infesting the stem in severely infected palms. Sporophores of the fungus, *G. lucidum* appear at the base of the affected palms prior to wilting or just after the death of the palm.

Crown: The leaflets exhibit wilting symptoms and outer one or two whorls of leaves turn yellow. Later, they exhibit light to moderate browning followed by drooping and drying. As the disease advances, the remaining leaves also droop down in quick succession and the spindle alone remains. The first external symptoms are flaccidity and folding of leaflets, chlorosis and bronzing of lower whorl of leaves. Under prolonged infection, the outer leaves falloff one by one, leaving only the spindle with a few unhealthy leaves around. The spindle leaves which emerge subsequently are reduced in size and do not unfold properly. Later stem shrivels and dries up. In some cases leaves break off near the base along the midrib. In advanced stages, all the leaves drop off leaving very thin decapitated stem. As the disease progresses, number of flowers, number of buttons reduces and normal development is arrested leading to button shedding. The leaves droop down resulting in hanging down of the subtended bunches. Most of the palms bear profusely, just prior to and at the time of initiation of symptoms.

Epidemology: The disease is mostly prevalent in sandy soils and where coconut gardens are raised under rain fed conditions. Lack of soil moisture during summer months, presence of old infected stumps in the garden, injury to roots and non adoption of recommended cultural practices favored the disease spread. The spread of the disease was found to be negatively correlated with total rainfall and number of rainy days.

Management

In view of the soil borne nature, survival and spread of the pathogen and losses caused by this disease, the application of biocontrol agents in the field is the only option to suppress the pathogen and further it helps in recovery of plants as it will be evident through increased foliage, decreased disease load and increased yield per tree. The mortality of the plants can also be minimized and managed with commercial biocontrol agents.

Bio-control based integrated disease management package against basal stem rot disease of coconut

The identified prophylactic as well as curative measures and indicator plants were developed as a bio control based integrated disease management package against basal stem rot disease of coconut. The measures of the bio control based integrated disease management package involves the following steps:

- ☆ Application of recommended dose of fertilizers to the coconut palms every year.
- ☆ Drip or basin method of irrigation is to be followed while irrigating the fields.
- ☆ Frequent watering or irrigation is to be applied to the palms especially during summer months.
- ☆ While irrigation, care should be taken to avoid flow of water from diseased trees to other healthy trees.
- ☆ Injury or damage to roots and pruning and cutting of the roots needs to be avoided.
- ☆ Raising and ploughing in situ of green manure crops like Sunhemp and Sesbania was followed for increasing the soil organic matter and antagonistic microflora.
- ☆ Sowing of indicator plants, Red gram and Bengal gram seeds in the tree basins for symptom development thereby early detection of the disease.
- ☆ Red gram plants show bark splitting symptoms as the identification mark for basal stem rot disease.
- ☆ Bengal gram plants shows withering, yellowing and drying of lower set of leaves followed by upper leaves as the identification mark of basal stem rot disease or *Ganoderma* wilt disease of coconut.
- ☆ Frequent observation and detection of the disease symptom on coconut stem at early stage is utmost important in managing the disease by any measure. The treatment is effective when the disease is in its early stage.
- ☆ Uprooting and destruction of diseased and dead palms along with the roots.
- ☆ Isolation of diseased trees from healthy palms by digging isolation trenches of 1m depth and 0.5m depth.

- ☆ As a prophylactic measure, application of 50g of *Trichoderma reesei* in combination with one kg of neem cake to the all healthy palms in the diseased garden keeps the disease away.
- ☆ As a curative measure, soil application of *Trichoderma reseei* @ 125 gm + *Pseudomonas fluorescens* @ 125 gm + neem cake@5 kg /palm/year to the diseased palm is effective.
- ☆ Clean cultivation and cultural practices needs to be followed

Stem Bleeding Disease

Stem bleeding disease of coconut is another debilitating disease of coconut and is prevalent in all coconut areas of tropics. The disease is caused by fungal pathogen, *Thielaviopsis paradoxa* (Deseynes) in all the soil types. Percentage disease incidence of stem bleeding was reported up to 15% in Andhra Pradesh (Srinivasulu *et al.,* 2005). Disease severity is more expressive in dwarf and soft tissued varities like Ganga Bondam[Gowthami Ganga]. The conventional varieties like east coast tall in Andhra Pradesh and west coast tall in Karnataka and Kerala are comparatively tolerant to the disease.

Symptoms: The disease is characterized by development of dark brown patches appearing at the basal portion of the trunk. A dark reddish brown liquid exudes from the longitudinal growth cracks present on the stem bark and form irregular streaks of exudation. These streaks may coalesce and form larger lesions. No oozing is seen from old lesions. The exudates eventually dry up to form black encrustations with brownish orange margins. The tissues beneath the discolored patch show decay. As the decay progresses, the tissues become black and fibrous. As a result of this, cavities are formed from which liquid comes out, when the bark is pressed. Maximum symptoms of bleeding during the rainy season and the least in summer months was noticed and disease progress was faster during months of high humidity (July to November months) and moderate temperature. The bleeding patches spread spirally about halfway up the stem and sometimes reach the crown and cause the death of palms.

Epidemiology: *Thielaviopsis paradoxa* (Deseynes) is a weak pathogen The fungus is widely distributed in the tropics. It attacks coconut palm, sugarcane, pine apple etc. It may gain entry through growth cracks or injuries made by boring insects or with human interference. The spores of the pathogen can also be disseminated by water, air currents or by insect vectors. The growth cracks that are developed after sudden rains following prolonged dry period may pave entry of the pathogen. Trash burning at the base also facilitates pathogen entry of the pathogen due to development of cracks. Drainage, hard lateritic pan formation leading to crippled root growth, imbalanced mineral nutrition are the predisposing factors of the disease. The micronutrient Chlorine play important role in the manifestation of bleeding symptoms. Excessive salinity with high sodium during summer leads to stem bleeding disease in coconut.

Bio-control based integrated disease management package against stem bleeding disease

- ☆ Injury or damage to the stem, especially while tractor ploughing should be avoided.
- ☆ Recommended dose of organic (50 kg FYM or green leaf manure or compost) and inorganic fertilizers (NPK 500 g: 320 g: 1200 g) may be applied per palm/year.
- ☆ Frequent watering / irrigation should be done during summer months. While irrigating, care should be taken to avoid flow of water from diseased trees to others. Basin system of irrigation to individual palms should be adopted.
- ☆ Good drainage facilities may be provided in the garden.
- ☆ Raising and ploughing insitu of green manure crops like Sunn Hemp and Sesbania is advised to increase soil organic matter and antagonistic microflora.
- ☆ *Trichoderma* Coir pith cake application on the stem bleeding patch and covering with leaf material and tagging is quickly cures the disease even in all seasons. Soil application of *Trichoderma reseei* @ 125 gm + *Pseudomonas fluorescens* @ 125 gm + neem cake@5 kg /palm/year to the diseased palm is advisable to suppress the soil borne inoculum.
- ☆ Smearing of talc powder paste of *Trichoderma* spp on the bleeding patches on the stem (paste can be prepared by adding 5 ml of water to 10 g of talc powder) is also effective in non rainy seasons.
- ☆ Soil application of 50 g of talc powder formulation of *Trichoderma* spp *viz., Trichoderma viride / T.harzianum / T.hamatum* in combination with 5 kg neem cake / palm /year to all the palms in diseases garden is advisable as a prophylactic measure.

Bud Rot Disease

Phytophthora palmivora is a ubiquitous pathogen that causes a variety of illnesses in a variety of plants. Bud rot in coconut caused by *Phytopthora palmivora* is a fatal disease characterized by the rotting of the terminal bud and surrounding tissues. Even though it affects the palms of all ages, young palms in low lying and moist situations are more susceptible to the disease. It is generally a sporadic disease, but sometimes appears in epidemics. Bud rot disease incidence on coconut recorded up to 15% in Andhra Pradesh.

Symptoms in Spindle: The first visible symptom is the withering of the spindle marked by pale colour. The spear leaf or spindle turns brown and bends over. Basal tissues of the leaf rot quickly and can be easily separated from the crown. Spindle withers and droops down and one by one, the inner leaves also fall away, leaving only fully matured leaves in the crown. A foul smell is emitted by the rotting tissue.

The disease may spread to older, adjacent leaves and spathes, producing a dead centre with a fringe of living leaves.

Symptoms on Nuts: The fungus also infects nuts, causing decay of immature nuts and their fall during the rainy season. Water soaked grayish green area develops at the stalk end of the nuts against dark green healthy area around. The lesions later turn brown and become sunken due to decay of underlying tissues. The rot extends into the husk and sometimes deep into the endosperm cavity. Infected nuts show brown to black necrotic areas with a yellow border developing on the surface. Internally, they have a mottled appearance. Young nuts are highly susceptible and fail to mature and then fall off from the tree.

Epidemiology: The disease is rampant during the monsoon when the atmospheric temperature is low and the humidity is high. *Phytophthora palmivora* Butler EJ, et al. is a cosmopolitan pathogen with a wide host range, including some very important economic crops such as cocoa, papaya, black pepper, rubber, coconut, and citrus etc. *Phytophthora* [water moulds] survive in the soil and plant debris, probably in a dormant state as thick-walled spores called "chlamydospores". These germinate producing spores called "sporangia" that are spread by wind, wind-driven rain and possibly by insects.

Bio-control based integrated disease management package against bud rot disease

- ☆ Bud rot disease can be effectively managed by adopting integrated management practices
- ☆ Dwarf palms are more sensitive to bud rot disease. Local ecotypes are generally more tolerant to *P. palmivora* than introduced ecotypes
- ☆ In nursery, application of *Pseudomonas fluorescens* [or] *Trichoderma reesei* [or] *P. fluorescens + T. reesei* @ 5gm per coconut seedling at Spindle region is effective.
- ☆ In the endemic areas preventive measures like Crown cleaning followed by whorl application of *Pseudomonas fluorescens* [or] *T. reesei* [or] *P. fluorescens* + *T. reesei* @ 100 gm / palm/year around the base of the spindle in the innermost leaf axils before the onset of monsoon is a very effective prophylactic treatment for the 3 years old young palms and increase the dosage use up to 200 gm for the 5 years and above aged palms is effective.
- ☆ In the early stage of the disease, remove the spindle leaf by pulling it out and cut and remove the infected tissues completely. The wound should be treated with Bordeaux paste 10 %.
- ☆ The diseased tissues should be burnt after their removal.
- ☆ Rhinoceros beetle attack predisposes the palm to bud rot infection. Hence prophylactic measures to prevent beetle infestation have to be undertaken in bud rot endemic areas

- ☆ Talc based application of *T.reesei* @ 5 g/seedling of coconut at spindle region is found to be effective for the management of bud rot disease in nursery seedlings.

Leaf Blight

Leaf blight disease of coconut caused by *Lasiodiplodia theobromae* (Syn: *Botryodiplodia theobromae*) (Pat.) Griffon and Maubl is an emerging problem in Andhra Pradesh and Tamil Nadu. For the first time in Andhra Pradesh recorded the coconut leaf blight disease incidence caused by *Lasidodiplodia theobromae* in Jangareddy gudem village of West Godavari during 2019-2020. In recent years, the disease is spreading at a faster rate in Eluru District and surrounding areas.

Symptoms on Leaf: The initial symptoms of the disease are confined to the outer whorl leaves and spreads to the inner leaves. Spindle leaves do not show up the disease during low infestation. The initiation of symptoms is usually from the tip of the leaflets as water soaked lesions. Occasionally it also initiates from the margin or any other part of the leaflet. The lesions gradually enlarge and coalesce resulting blighting followed by progressive drying of a larger portion of the frond. The infected tissues of leaflet turn brown to grey. Usually the spread of the disease is not restricted by the midrib of the leaflets. The midrib is also infected and dried. As the disease advances the size of the leaves is reduced and production of female flowers is reduced. Affected leaflets start drying from the tip downwards and exhibit a charred or burnt appearance.

Symptoms on Nut: Sometimes the pathogen also infects the buttons and immature nuts, produces characteristic symptoms and led to dropping. Dark grey to brown lesions with wavy to undulated margins appear from the apex of the nuts. The fungus entered in to the kernel through functional pore, resulting in to decay of the endosperm. The affected nuts were desiccated, shrunken, deformed and dropped prematurely and resulting in nut yield loss up to 10 to 25%.

Epidemiology: Under favorable conditions the disease spreads fast and results in blighting of a large area of the foliage. Usually the disease is severe during the summer months. The fungus over-winters as pycnidia on the debris. The pycnidium produces two-celled, dark brown, striated conidia. The conidia are then dispersed by wind and rain splash. Disease develops when conidia land on fresh leaf and germinate with thin film of water. The snow fall resulted moisture film may also help the germination of conidia during summer. The conidia germinate and infects through stomata and start causing damage .

Integrated Management of *Lasiodiplodia* leaf blight

- ☆ Removal and destruction of severely affected fronds (once in 3 months)
- ☆ Basal application of Dr.YSRHU - Coconut IDM Special- *Pseudomonas fluorescens* + *Trichoderma* @ 200g along with 50 kg FYM/palm / year as a prophylactic measure.

- ☆ In severe cases foliar spray of Carbendazim (0.1%) followed by Hexaconazole (0.1%) after15 days of 1st spray as a curative measure.

Standardized protocol for mass multiplication of bioagents

Protocol for mass multiplication of fungal bioagents (*Trichoderma* spp)

- ☆ The culture of *Trichoderma* spp was subcultured and allowed to grow on petri dishes containing potato dextrose agar medium for a week.
- ☆ A piece of week-old fungal mat from the PDA plate was then inoculated into potato dextrose broth in a conical flask / fermentor and incubated at room temperature for another week.
- ☆ The mycelial mat thus obtained was homogenized and mixed with talc powder at 1: 2 ratio.
- ☆ Carboxy Methyl Cellulose at the rate of 5g per kg was added to the talc powder and the mixture was allowed to shade dry at room temperature
- ☆ Finely macerated powder packed and labeled.

Protocol for mass multiplication of bacterial bioagent (*P. fluorescens*)

- ☆ The *P. fluorescens* bacterial culture was grown on petri dishes containing Kings B medium for overnight periods.
- ☆ A loopful of fresh bacterial culture was then inoculated to Kings B broth and incubated at room temperature for 4 days.
- ☆ The liquid culture was then mixed with talc powder at 1: 2 ratio.
- ☆ Carboxy Methyl Cellulose at the rate of 5 g per kg was added to the talc powder and allowed to shade dry at room temperature.
- ☆ Finely macerated powder packed and labeled.

References

Alouw, J. C., Rejesus, B. M. and Ocampo, V. R. 2005. Biology of the coconut spike moth, Ananthakrishnan, T. N. 2009. Invasive insects in agriculture, medicine and forestry.

Anonymous. 2018. Pest alert - A new Invasive Bondar's Nesting Whitefly reported on coconut from Kerala, India. CPCRI.

Anonymous. 2019. http://coconutboard.nic.in/Statistics.aspx

Baozhu, Z., Chaojun, L. V. and Weiquan, Q. 2017. Effect of temperature on the population Growth of *Tirathaba rufivena* (Lepidoptera: Pyralidae) on *Areca catechu* (Arecaceae). *Florida Entomologist*. **100**: 578-582.

Basri, M. W., Norman, K. and Hamdan, A. B. 1995. Natural Enemies of the Bagworm, *Metisa Plana* Walker (Lepidoptera: Psychidae) and their impact on Host Population Regulation. *Crop Protection*. **14** (8): 637-645 .

Bouèek, Z., 1988. Australasian Chalcidoidea (Hymenoptera). A Biosystematic Revision of Genera of Fourteen Families, with a reclassification of species. *CAB International, Wallingford, Oxon, U.K.*, pp 701.

Cantone, F. A. and Vandenberg, J. D. 1998. Intraspecific diversity in *Paecilomyces fumosoroseus. Mycol Res.* 102: 209–215.

Chalapathirao, N. B. V., Nischala, A., Ramanandam, G. and Maheswarappa, H. P. 2018b. Biological suppression of coconut black headed caterpillar *Opisina arenosella* outbreak in East Godavari district of Andhra Pradesh – eco friendly technology. *Current Science.* **115** (8): 1588-1594.

ChalapathiRao, N. B. V., Snehalatharani, A., Nischala, A., Ramanandam, G. and Maheswarappa. H. P. 2018a. Management of rhinoceros beetle (*Oryctes rhinoceros* L.) by biological suppression with *Oryctes* baculovirus in Andhra Pradesh. *Journal of Plantation Crops.* **46**(2): 124-127.

Cock, M. J. W. and Perera, P. A. C. R. 1987. Biological control of *Opisina arenosella* Walker (Lepidoptera: Oecophoridae). Coconut Research Institute of Sri Lanka. *B.N.I.* 8: 283-309.

Cock, M. J. W., Gallego, C. N. and Godfray, H. C. J. 1987. Biological control of *Tirathaba rufivena* in the Philippines. Paper prepared for the ACIAR workshop *on* Biological Control held in Tonga, 17-25 October 1987.

Evans, G. A. The whiteflies (Hemiptera: Aleyrodidae) of the world and their host plants and natural enemies. USDA-APHIS. 2008; (31 July 2013).

Francis, A.W, Stocks, I. C., Smith, T. R., Boughton, A. J., Mannion, C. M. and Osborne, L. S. Host plants and natural enemies of rugose spiraling whitefly (Hemiptera: Aleyrodidae) in Florida. *Florida Entomol.* 2016; 99(1): 150–153.

Frank, K. D. 1988. Impact of outdoor lighting on moths: an assessment. *Journal of the Lepidopterists Society.* **42**:63-93

Ganiger, P.C., Yeshwanth, H. M., Muralimohan, K., Vinay, N., Kumar, A. R. V. and Chandrashekara, K. 2018. Occurrence of the new invasive pest, fall armyworm, *Spodoptera frugiperda* (J.E. Smith) (Lepidoptera: Noctuidae), in the maize fields of Karnataka, India. *Current Science.*; **115**(4): 621-623.

Ghosh, S. M. and Abdurahiman, U. C. 1988, Biology of *Pediobius imbrues* (Hymenoptera: Eulophidae), a hyperparasite of *Opisina arenosella*, the black headed caterpillar of coconut. *Cocos.* **6**: 27-30.

Godfray, H. C. J. 1985. Mass rearing the Tachinid fly *Argyrophylax basifulva* a parasitoid of the greater coconut spike moth (*Tirathaba* spp.) (Lep: Pyralidae). *Entomophaga.* **30**(3): 211-215.

Hoenicka, H. and Fladung, M., , Biosafety in *Populus* spp. and other forest trees: from non-native species to taxa derived from traditional breeding and genetic engineering. *Trees: Structure and Function* 2006; 20(2): 131-144.

Howard, F. W. 2001. Insect pests of palms and their control. *Pesticide outlook*. 12: 240–244.

Hoy, M. A., Singh, R. and Rogers, M. E. 2010. Evaluations of a novel isolate of *Isaria fumosorosea* for control of the asian citrus psyllid, *Diaphorina citri* (Hemiptera: Psyllidae). *Fla Entomol.* 93: 24–32.

Idrus, M. S., Hassan, A., Abdul Hafiz, A. M. and Aiman, H. J. 2016. Fruit bunch moth *Tirathaba* sp. (Lepidoptera: Pyralidae) infestation in oil palm planted on peat soil: biology and ecology. In: Proceedings of the Second Entomology Postgraduate Symposium (ENTOPOST 2016). Bangi, Selangor, Malaysia, 29-30 March.

Ismail, B. S., Maslinda, M. and Zuriati, Z. 2012. Effects of temperature, soil moisture content and soil type on the degradation of cypermethrin in two types of Malaysian agricultural soils. *World Applied Sciences Journal.* **17**(4): 428-432.

Jayanthi, P. K. D. and Verghese, A. 2009. Phototactic response of sapota seed borer, *Trymalitis margarias* Meyrick, a clue for integrated pest management. *Pest management in Horticultural ecosystem*. **15**(1):68-69.

Kalidas, P. 2002. Incidence of leaf eating caterpillars on oil palm and strategies for their management. In: *Placrosym XV*, Mysore, India, 10-13 December, 2002.

Lim, K. H. 2012. Integrated pest management of *Tirathaba* bunch moth on oil palm planted on peat. *The Planter.* **88**(1031): 97-104.

Mani, M. Origin, introduction, distribution and management of the invasive spiralling whitefly *Aleurodicus dispersus* Russell in India. *Karnataka J Agric Sciences* 2010;**23**(1): 59–75.

Mani, M., Dinesh, M.S. and Krishnamoorthy, A. Presence of *Encarsia* spp. on spiralling whitefly *Aleurodicus dispersus* (Russell) in peninsular India. *Insect Environment* 2000;**6**, 100.

Mani, M., Krishnamoorthy, A and Venugopalan, R. Role of the aphelinid parasitoid *Encarsia guadeloupae* Viggiani in suppression of exotic spiralling whitefly *Aleurodicus dispersus* Russell on banana in India. *Biocontr. Sci. Tech.* 2004;**14**: 619–622.

Mannion, C. 2010. Rugose spiraling whitefly, a new whitefly in South Florida. Tropical Research and Education Center, University of Florida: pp 5. (17 January 2017).

Martin, J.H. 2004. The whiteflies of Belize (Hemiptera: Aleyrodidae) Part 1 - introduction and account of the subfamily Aleurodicinae Quaintance & Baker. *Zootaxa*. **681**: 1–119.

Muthiah, C. 2007. Estimation of yield loss caused by eriophyid mite on coconut. *Annals of Plant Protection Sciences*. **15**: 484–486.

Nair, M. R. G. K and Visalakshi, K. 1999. A monograph on crop pests of Kerala and their control, Kerala Agricultural University, Vellanikkara.

Patel, R. K., Salam, P. K. and Singh, B . 2018. New record of coconut spike moth (*Tirathaba rufivena* Walker) from Bastar tribal belt of Chhattisgarh. *Journal of Entomology and Zoology Studies.* **6**(1): 1117-1118.

Polaszek, A., Evans, G.A. and Bennett, F.D. *Encarsia* parasitoids of *Bemisia tabaci* (Hymenoptera: Aphelinidae, Homoptera: Aleyrodidae): a preliminary guide to identification. *Bull Entomol Res.*1992;**82**: 375–392.

Polaszek, A., Manzari, S and Quicke, D. L. J. Morphological and molecular taxonomic analysis of the *Encarsia meritoria* species-complex (Hymenoptera, Aphelinidae), parasitoids of whiteflies (Hemiptera, Aleyrodidae) of economic importance. *Zoologica Scripta* 2004;**33**(5): 403–421.

Poorani, J. and Thanigairaj, R. 2017. First report of *Encarsia dispersa* Polaszek (Hymenoptera: Aphelinidae) as a parasitoid of rugose spiralling whitefly, *Aleurodicus rugioperculatus* Martin (Hemiptera: Aleyrodidae), a recent invasive pest in India, with notes on its predators. *Journal of Biological Control.* **31**(1): 1-4.

Raghubanshi, A.S., Raj, L.C., Gaur, J.P. and Singh, J.S. 2005. Invasive alien species and biodiversity in India. *Curr. Sci.* **88**: 539-540.

Rajan, P., Josephrajkumar, A. and Sujatha, A. 2011. Gradient outbreak of coconut slug caterpillar *Macroplectra nararia* Moore in east coast of India. *Cord.* **27**(1):61-68.

Ramachandran, C. P., Ponnamma, Abdulla Koya, K. M. and Kurian, C. 1974. The coconut leaf eating caterpillar, *Nephantis serinopa* Meyrick, a review. *Philippines J. Cocon. Stud.* **4**: 9–17.

Ramani, S. Fortuitous introduction of an aphelinid parasitoid of the spiralling whitefly, *Aleurodicus dispersus* Russell (Homoptera: Aleyrodidae) into the Lakshadweep islands, with notes on host plants and other natural enemies. *Journal of Biological Control* 2000;**14**, 55-60.

Ramani, S., Poorani, J. and Bhumannavar, B.S. Spiralling whitefly, *Aleurodicus dispersus* Russell (Homoptera: Aleyrodidae) in India. *Biocontr News Infor.* 2002; **23**(2): 55N–62N.

Ramkumar, Muralimohan, K., Kiranmayi, L. and Srinivasa, Y. B. 2006. Discrete generation cycles in the tropical moth *Opisina arenosella*. *Current Science.* **91**: 811–816.

Selvaraj, K., Sundararaj, R., Venkatesan, T., Ballal, C. R., Jalali, S. K., Gupta, A. and Mrudula, H. K. 2016. Potential natural enemies of the invasive rugose spiraling whitefly, *Aleurodicus rugioperculatus* Martin in India. *Journal of Biological Control.* **30**(4): 236-239.

Selvaraj, K., Sundararaj, R., Venkatesan, T., Ballal, C. R., Jalali, S. K., Gupta, A. and Mrudula, H. K. 2016. Potential natural enemies of the invasive rugose spiraling whitefly, *Aleurodicus rugioperculatus* Martin in India. *Journal of Biological Control.* **30**(4): 236-239.

Shanas, S., J. Job, T. Joseph and Krishnan, G. A. 2016. First report of the invasive rugose spiralling whitefly, *Aleurodicus rugioperculatus* Martin (Hemiptera: Aleyrodidae) from the Old World. *Entomon.* **41**(4): 365 – 368.

Shivasankar, T., Annadurai. R.S., Srinivas, M. and Preeti, G. 2007. Control of black headed caterpillar *Opisina arenosella* Walker by systematic application of 'Soluneem' a new water soluble neem insecticide formulation. *Current Science.* **8**: 176–179.

Smith, P. 1993. Control of *Bemisia tabaci* and the potential of *Paecilomyces fumosoroseus* as a biopesticide. *Biocontrol News Inf.* 14: 71–78.

SnehalathaRani, A., Chalapathirao, N. B. V. and Emmanuel, N. 2012. New Record of *Paecilomyces lilacinus* (Deuteromycota: Hyphomycetes) as an entomopathogenic fungi on slug caterpillar of coconut. *Insect Environment.* **17**(4): 151-152.

Stocks, I. 2012. Rugose spiraling whitefly host plants. Florida Department of Agriculture and Consumer Services (FDACS), Division of Plant Industry (DPI).

Stocks, I. and Hodges, G. Pest Alert- DACS-P-01745. 2012. Florida Department of Agriculture and Consumer Services, Division of Plant Industry. pp. 6. (17 January 2017).

Sujatha, A. and Chalapathirao, N. B. V. 2004. Natural occurrence of bioagents (*Oryctes baculovirus* and *Metarhizium anisopliae*) of rhinoceros beetle in Andhra Pradesh. *Insect Environment.***10** (2): 69-70.

Sujatha, A., Chalam, M.S.V., Vijaylakshmi, P. and Gautam, B. 2008. Coconut slug caterpillar *Macroplectra nararia* – Out break in Godavari districts of Andhra Pradesh. *Indian Coconut Journal.* **38**(11):2-4.

Sujatha, A., Emmanuel, N. and ArulRaj, S. 2011. Light trap - induced suppression of coconut slug caterpillar, *Macroplectra nararia* Moore menace in East coast of India, *Journal of plantation Crops.* **39**(3): 390 – 395.

Sujatha, A., Nair, C. P. R. and Chalapathirao N. B. V. 2004. Integrated Pest Management Technology for Rhinoceros Beetle, *Oryctes rhinoceros* L. In Coconut Plantations. *The Andhra Agricultural Journal* ***50*** *(spl) – Golden jubilee special issue*: 254-256.

Sundararaj, R. and Selvaraj K. Invasion of rugose spiralling whitefly, *Aleurodicus rugioperculatus.* Martin (Hemiptera: Aleyrodidae): a potential threat to coconut in India. *Phytoparasitica*, 2017; DOI:10.1007/s12600-017-0567-0.

Taravati, S., Mannion, C., Glenn, H. and Osborne, L. Natural enemies of rugose spiraling whitefly, *Aleurodicus rugioperculatus* Martin (Insecta: Hemiptera: Aleyrodidae) in the south Florida landscape. ENY- 870. University of Florida, IFAS Extension. 2013; Available from: http://edis.ifas.ufl.edu/pdffiles/IN/IN100400. pdf.

Thangjam, B., Khan, M. A., Pandey, S and Manhanvi, K. 2013. A new species of genus Pediobius Walker (Hymenoptera: Eulophidae) with new records of eulophid parasitoids from Uttarakhand, India. *Pantnagar Journal of Research.* **11**(1): 29.

Tirathaba fructivora (Meyr.) (Lepidoptera: Pyralidae). *Philippines Entomologist.* **19** (1): 84-93.

Zelazny, B.1985. Susceptibility of two coconut pests, *Oryctes rhinoceros* (Col.: Scarabaeidae) and *Tirathaba rufivena* (Lep.: *Pyralidae*) to the entomoparasitic nematode *Steinernema feltiae (=Neoaplleactana carpocapsae). Entomophaga.* **30**(2): 121-124.

Zimmermann, G. 2008. The entomopathogenic fungi *Isaria farinose* (formerly *Paecilomyces farinosus*) and the *Isaria fumosorosea* species complex (formerly *Paecilomyces fumosoroseus*): biology, ecology and use in biological control. *Biocontrol Science and Technology*. **18**(9): 865-901.

Coconut black headed caterpillar larvae

Bracon hebetor

Goniozus nephantidis

Coconut black headed caterpillar infested garden

Bracon brevicornis

Coconut slug caterpillar infested leaf

Pediobius imbrues

Rhinoceros beetle damage

Red palm weevil damage

***Apertochrysa astur* eggs**

***Apertochrysa astur* grub**

Fig1: Predator, *Cryptolaemus*,a ladybird beetle feeding on mealybug species. (Courtesy: M. J. Raupp, Cornell University)

Fig 2: Parasitoid *Encarsia* against whitefly. (Courtesy: M. Hoddle, Cornell University)

SYMPTOMS MAJOR DISEASES OF COCONUT

Stem bleeding disease

Basal stem rot (Ganoderma wilt) disease

Bud rot disease

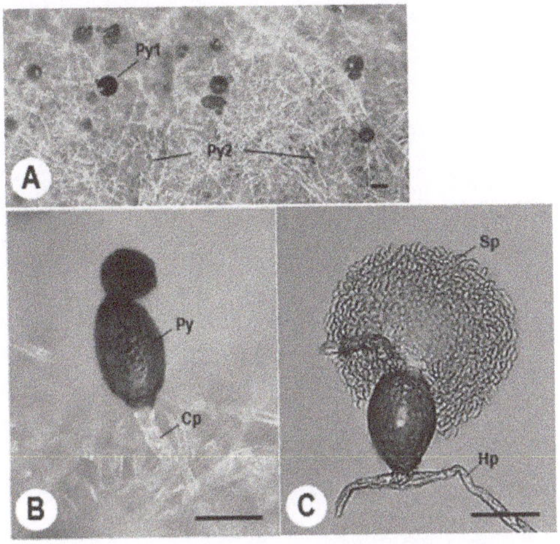

Fig 3: Ampelomyces strain against Powdery mildew pathogen (Courtesy: Nemeth et al., 2021)

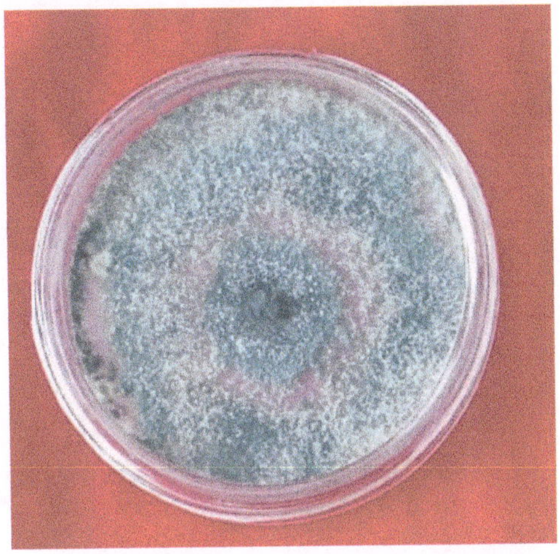

Fig 4: Culture of *Trichodema*, a very promising biological control agent

Chapter-5

INTEGRATED MANAGEMENT OF INSECT PESTS AND DISEASES IN OIL PALM WITH EMPHASIS ON BIOLOGICAL CONTROL

V. Anoosha[1]*, G. Koteswara Rao*[1]*, M. Tirupathi Reddy*[2]*, B.K.M. Lakshmi*[3]

[1]*Dr YSRHU-Horticultural Research Station*

Dr. YSR Horticultural University

Ambajipeta, Dr. B. R. Ambedkar Konaseema District, A.P.

[2]*Dr YSRHU-Horticultural Research Station*

Dr. YSR Horticultural University

Vijayarai, Eluru District, A.P.

[3]*Dr YSRHU-Mango Research Station*

Dr. YSR Horticultural University

Nuzvid, Eluru District, A.P.

Introduction

Oil palm (*Elaeis guineensis*; Family: Arecaceae) is grown for its oil which is used in cooking and in industry and it is referred as small holders irrigated crop and it plays vital role in countrys economoy too . The edible oil of *Elaeis guineensis* oil palm is crucial in filling the demand gap and meets the growing need for edible oil worldwide. Oil palm trees are unbranched with a long stout single stem, or trunk, terminating in a crown of 7–100 leaves. The leaves are pinnate and can reach 3–5 m (9.8–16.4 ft) in length. The tree produces large, spherical red fruits in bunches. Up to 200 fruits can be produced per bunch and the oil is extracted from the pulp and kernel. Oil palm can reach heights of 20–30 m (65.6–98.4 ft) and has an economic lifespan of 25–30 years, at which point they become too tall to be managed efficiently and are cut down. Left alone, oil palm has been known to live for periods up to 200 years. Oil palm is the richest source for vegetable oil production with a capacity of 4-6 tons of oil per ha per year. It is the most sustainable crop to feed the hungry mouths of the world as it is recognized universally as the most efficient, effective and highest yielding form of edible oil production. The eco friendliness is manifested as it is a perennial plantation crop, maintains its green canopy throughout 30 years of its economic life and does not cause soil erosion, river siltation, etc.

Palm oil is widely used as cooking oil, as an ingredient in many processed foods and as a substitute for butter. Palm oil can be used to produce biodiesel. Palm oil is also used as a binding agent in cosmetics. Oil palm is typically grown in tropical lowland regions over large areas. Harvesting of oil palms generally begins 2.5 years after the trees are planted in the field. Harvesting takes place throughout the year, usually at 2–3 week intervals, as the bunches ripen. Bunches are harvested by cutting from the tree using a sharp knife or sickle.

This multipurpose plantation crop that is extensively grown in the humid tropics across Asia, Africa and the Americas for palm oil and other uses such as sauces, soap, wine, fertilizer (ashes), roofing (leaves), building material (trunk), medicines (roots), and ornamental purposes (Murphy *et al.* 2021). About 40% of all traded vegetable oil globally is palm oil (Murphy *et al.* 2021), and this oleaginous species has the potential to produce 4–20 times more oil per hectare of land compared to other oil crops. As such, oil palm production has rapidly grown more than 35-fold from only 2 million tonnes in 1970 to 71 million tonnes in 2018. There are more than 140 palm oil importing countries of which India, China and the United States of America lead the market with more than nine, five and one million tonnes of palm oil imports in 2017, respectively (Ritchie and Roser, 2022).

Pests and diseases are one of the factors which hinders the growth and development of oil palm which results in yield loss .Although all precautions have been taken to minimize the introduction of pest and diseases along with seed sprouts from importing countries, some pests and diseases are still found to invade crops and cause production losses. Most of these pests and diseases have been found to come from other Arecaceae/Palmaceae palms such as oil palm, palmyrah and betel nut that may be commonly found around oil palm fields. In this chapter various pests and diseases that affecting the oil palm crop and their integrated management is dicussed .

Definition of IPM

"IPM is a sustainable approach to managing pests by combining biological, cultural, physical and chemical tools in a way that minimizes economic, health, and environmental risks."

Aims of IPM

- ☆ Reduce the use of synthetic organic pesticides
- ☆ That are environmentally sound
- ☆ Pest minimal risk of human health
- ☆ Re-useable return on investment
- ☆ Provide consumable safe food

Principles of IPM

- ☆ Identification of key pests and beneficial organisms
- ☆ Defining the management unit, the Agro-ecosystem
- ☆ Development of management strategy
- ☆ Establishment of Economic thresholds (loss & risks)
- ☆ Development of assessment techniques
- ☆ Evolving description of predictive pest models

Tools of IPM

Monitoring

Keep tracks of the pests and their potential damage. This provides knowledge about the current pests and crop situation and is helpful in selecting the best possible combinations of the pest management methods.

Pest resistant varieties

Breeding for pest resistance is a continuous process. These are bred and selected when available in order to protect against key pests.

Cultural pest control

It includes crop production practices that make crop environment less susceptible to pests. Crop rotation, cover crop, row and plant spacing, planting and harvesting dates, destruction of old crop debris are a few examples. Cultural controls are based on pest biology and development.

Mechanical control

These are based on the knowledge of pest behaviour. Hand picking, installation of bird perches, mulching and installation of traps are a few examples.

Biological control

These include augmentation and conservation of natural enemies of pests such as insect predators, parasitoids, and pathogen and weed feeders. In IPM programmes, native natural enemy populations are conserved and non-native agents are released with utmost caution.

Chemical control

Pesticides are used to keep the pest population below economically damaging levels when the pests cannot be controlled by other means. It is applied only when the pest's damaging capacity is nearing to the threshold.

Biological practices

Biological control of insect pests and diseases through biological means is most important component of IPM. In broader sense, biocontrol is use of living organisms to control unwanted living organisms (pests). In other words, deliberate use of parasitoids, predators and pathogens to maintain pest population at level blow those causing economic loss either by introducing a new bioagent into the environment of pest or by increasing effectiveness of those already preset in the field.

Parasitoids: These are the organisms which lay eggs in or on the bodies of their hosts and complete their life cycles on host bodies as a result of which hosts die. A parasitoid may be of different type depending on the host developmental stage in or on which it completes its life cycle. For example, egg, larval, pupal, adult, egg-larval and larval pupal parasitoids. Examples are different species of *Trichogramma, Apanteles, Bracon, Chelonus, Brachymeria, Pseudogonotopus* etc.

Predators: These are free living organisms which prey upon other organisms for their food. Examples are different species of spiders, dragon flies, damsel flies, lady bird beetles, *Chrysopa* species, birds etc.

Pathogens: These are micro-organisims which infest and cause diseases in their hosts as a result of which hosts are killed. Major groups of pathogens are fungi, viruses and bacteria. Some nematodes also cause diseases in some insect pests. Important examples of fungi are different species of *Hirsutella, Beauveria, Nomurae and Metarhizium* which have been reported to infest and kill large number of insects (upto 90%) in the fields. Among viruses, most important examples are of nuclear polyhedrosis virus (NPV) and granulosis viruses. Outbreak of viruses in armyworms, cut worms, leaf folders, hairy caterpillars and plant hoppers have been reported many times. Among bacteria, *Bacillus thuringiensis (B.t.)* and *B. popillae* are very common examples.

Biocontrol Practices

Diseases of pests can be mass multiplied in the laboratory at a low cost in liquid or powdered formulations that can be sprayed like ordinary chemical pesticides. These formulations are known as biopesticides. The different types of biocontrol practices are grouped as under.

a. Introduction

In this process, a new species of bioagent is introduced into a locality for its establishment against its host. This is done only after thorough laboratory examination and field trials for its efficacy.

b. Augmentation

In this process, the population of natural enemies already present in the area is increased by releasing either laboratory reared or field collected bioagents of same species in such number as would require to suppress the pest population in that area.

c. Conservation

This is most important component of biological control and plays a major role in pest suppression. In this process, natural enemies present in the nature are protected from being killed. The different practices required to protect the natural enemies are as below.

- ☆ Collection of parasitised egg masses and placing them in bamboo cage-cum-bird perchers for allowing emergence of parasitoids and withholding of pest larvae.
- ☆ Educating farmers through field days, radios & TV to differentiate pests and defenders and sparing the defenders during field sprays.
- ☆ Chemical spray should be adopted as last resort and that too after observing pest defender ratio and economic threshold level (ETL).
- ☆ Use of broad spectrum pesticides should be avoided.
- ☆ Only selective and relatively environmental friendly (REF) pesticides should be used where necessary.
- ☆ As far as possible strip or spot application of pesticides be carried out.
- ☆ Adjustment of time of sowing and harvesting to avoid the peak season of pest attack.
- ☆ Growing trap crop on the borders of main fields before the actual sowing of crop to trap pest and develop natural enemies.
- ☆ Root dip/seedling treatment for gall midge prone area.
- ☆ Crop rotation and inter-cropping also help in conservation of defenders.
- ☆ Recommended dose and concentration of pesticides should be used.

Major Insect Pests and Diseases occurring in Oil Palm

Insect Pests in Oil palm

In Oil palm as many as 55 insect pests are known to damage in India, of which the major pests are discussed.

- ☆ Rhinoceros Beetle (*Oryctes rhinoceros* Linn)
- ☆ Red palm weevil (*Rhynchophorus ferrugineus*)
- ☆ Mealy bugs (*Dysmicoccus brevipes*)
- ☆ Rugose spiraling whitefly (*Aleurodicus rugioperculatus*)

☆ Spindle bug - *Mircarvalhoia (=Carvalhoia) arecae*

Rhinoceros Beetle (*Oryctes rhinoceros*)

Nature of damage and Symptoms

Oil palm palms aged between 3-10 years are more prone to attack of rhinoceros beetles. The affected fronds (leaves) when fully opened shows geometric or V-shaped cuts. The formation of round to oblong holes on the unopened spathes and spindle leaves. Repeated infestation to the growing point may lead to death of the seedlings. Injured portion may attract red palm weevil and fungal infection. Cyclone affected oil palm gardens, Vermicompost pits, decomposing coir pith and FYM heaps are breeding sites of Rhinoceros beetle.

Integrated pest management (IPM) practices for rhinoceros beetle

☆ **Sanitation:** Remove and destroy all possible decaying debris and dead palms from the plantation since this act as prolific breeding grounds for the beetle.

☆ **Prophylactic leaf axil filling:** To protect the young palms from rhinoceros beetle attack, the innermost 2-3 leaf axils are to be filled with a mixture of neem seed kernel powder 100g + fine sand (150g) per palm during May, September and December.

☆ **Biological management with the entomo pathogen *Metarhizium anisopliae*:** This fungus can be mass cultured in rhinoceros grubs or broken maize grain or oil palm water or on cassava chips and rice bran supplemented with a nitrogen source. Periodical spraying of fungus on manure heaps is to be done during monsoon season. The fungus should be inoculated @ 5×10^{11} spores/ m^3 of the breeding material.

☆ Placing of Napthalene balls @ 12 gm per palm in leaf axils will also deter beetles.

☆ **Traps for attraction:** Establishment of rhinolure traps @ 1 trap/ha. Will help in effective trapping of adult beetles.

☆ Regular monitoring for any damage in the spindle region of the juvenile palm.

Red palm weevil (*Rhynchophorus ferrugineus*)

The Infested oil palm shows yellowing and later wilting of leaves takes place. A Pencil size circular hole on the trunk and oozing of brownish viscous fluid from the trunk. Presence of chewed up fibrous material in the leaf axils.

Integrated pest management (IPM) practices for Red palm weevil

As far as possible avoid making any cuts causing injuries to the palm through agricultural tools and implements as these will attract weevil for egg laying. Chisel out affected trunk region and burn it and smear the wounded portion with coaltar .Afterwards fill the tunneled portion of trunk with cement and sand mixture to give strength to the palm.

The commercially available pheromone lure is hung inside from the lid of the bucket which is filled with 100 gm pineapple/sugarcane. The trap is hung at a height of 1 to 1.5 m above the ground level in the field @1 trap per hectare.

Mealy bugs (*Dysmicoccus brevipes*)

Symptoms

Flattened oval to round disc-like insect covered in waxy substance on tree branches; insects attract ants which may also be present; insect colony may also be associated with growth of sooty mold due to fungal colonization of sugary honeydew excreted by the insect; symptoms of direct insect damage not well documented. Insects have a wide host range; often tended by ants which farm them for their sugary honeydew secretions; transmit Cocoa swollen shoot virus.

Management

Mealy bugs can potentially be controlled by natural enemies such as lady beetles but are commonly controlled using chemicals; chemical pesticides may also decrease populations of natural enemies leading to mealy bug outbreaks.

Rugose spiraling whitefly (*Aleurodicus rugioperculatus*)

It is a serious pest of both nursery and main plantations. In India, the pest was first observed at Pollachi, Tamil Nadu during August 2016 (Sundararaj and Selvaraj, 2017) on oil palm. Due to its polyphagous nature, the pest has attained a key pest status in horticulturally important crops. Moreover, its especial preference to plantation crops made oil palm as a preferential crop, after oil palm causing substantial damage. Rugose spiraling whitefly is the major damaging species on oil palm whereas the bondar nesting whitefly is confined to some palms in a given plantation.

Biology: A female whitefly lays around 50 eggs in circular fashion and covers with white waxy material. After emergence, the nymphs settle nearby and continue to feed on the cell sap in groups. The nymphs also secrete white waxy material/ filaments from the lateral glads. The nymphal, pupal and adult periods lasts about 24, 12 and 23 days, respectively. The nymphal and adult stages of bondar nesting whitefly are located sparsely with individuals located centrally among the heavy secretion of waxy material.

Damage: It causes significant damage to both nursery seedlings and adult palms by sucking the sap from the underside of leaflets. Except the rainy season, the pest is seen throughout the year with its peak incidence from late winters to late summer months (November to May). After emergence from the egg, the newly hatched nymphs settle just few millimeters away from the egg, become sedentary and suck the sap from the leaves in a congregated manner. Besides this direct sap drain, formation of sooty mold on lower leaves also poses problem of reduction in photosynthetic ability (Kalidas, 2019).

Natural enemies: Under field conditions, the nymphal stages are parasitized by an aphelinid parasitoid, *Encarsia guadeloupae* (Hymenoptera: Aphelinidae) to the extent of 54%. The parasitized nymphs are readily identified by their dark colour and especially by the circular exit holes made by the adults. One green lace wing predator, *Pseudomallada astur* (Neuroptera: Chrysopidae) is also found to feed on all the stages of whiteflies. Besides, some spider species and coccinellids are also found feeding on the different life stages of invasive whiteflies.

Management: Due to the invasive nature of the pest, no insecticidal management is recommended against the pest especially on adult plantations. An entomopathogenic fungi, *Isaria fumasorosea* pfu 5 identified by ICAR-NBAIR, Bengaluru was found to be highly effective against the pest which canbe applied at 5ml/L of spray fluid along with 5% of any detergent powder. Installation of yellow sticky traps to trap the flying adults during dawn and dusk times also reduce the pest intensity. During early stages of infestation, application of neem oil @ 2.5ml/L along with detergent reducesthe severity and development of the pest. Conservation and release of mass produced predators or parasitoids is also an effective strategy for management of the pest.

Spindle bug *(Mircarvalhoia* (=Carvalhoia*) arecae)*

It is a major a pest of seedlings and young plants.

Biology: Eggs are laid singly between the leaflets of the spindle. The incubation period ranges from 10 to 12 days. Nymphs are light violet brown, greenish yellow with border of the body. There are five nymphal stages. The total nymphal duration of the bug ranges from 22 to 28 days. The light violet brown nymphs have greenish yellow border. Adults are bright red and black coloured. The bug completes its life cycle in 30-39 days. The pre-oviposition and oviposition period occupy 3 and 2 days respectively. The bug has the fecundity rate of 10-18 eggs. The adult male and female live for 20 and 25 days.

Damage: Both the nymphs and adults suck the sap from the tender spear leaves causing typical linear brown lesions. Spear leaves fail to open fully when the infestation is severe. The infested portionsdevelop necrotic patches, which turn brown and subsequently dry up.

The central portions of the necrotic patches drop off forming numerous holes on the leaves. Due to severe infestation the leaves are shredded and the palms become stunted (Nair and Daniel, 1982). Infestation is noticed throughout the year with the highest incidence during June and the lowest in February. The per cent infestation declines with the increase in the age of the palms (Dhileepan, 1992). Existence of low temperatures may be the critical factors for the pest incidence.

Management: The bugs are naturally suppressed by an entomopathogen, *Aspergillus candidus* Link. During the rainy season, coinciding with the peak period of its incidence (Dhileepan *et al.* 1990). The sachets are transferred over and again to the innermost leaf axils as and when new spindles emerge. The longevity of the sachet is about eight to ten months.

Major diseases of Oil palm

- ☆ Basal stem rot (*Ganoderma lucidum / G. applanatum*)
- ☆ Bud rot (*Phytophthora palmivora*)
- ☆ Stem Bleeding (*Thelaviopsis paradoxa*)
- ☆ Grey Leaf blight (*Pestalotia palmarum*)
- ☆ Bacterial bud rot (*Erwinia* spp.)
- ☆ Oil palm wilt *Fusarium oxysporum*

Basal Stem Rot or Ganoderma Wilt Disease (*Ganoderma lucidum / G. applanatum*)

Symptoms

- ☆ The disease first affects the underground roots causing root rot.
- ☆ Decay and death of the fine roots is the first underground symptom of the disease.
- ☆ Exudation of reddish brown viscous fluid from the basal portions of the stem is the first visible symptom of the disease in the affected palm.
- ☆ Outer whorls of leaves droop slowly and within a period of two years after the palm succumb to the disease, the entire crown collapses and the palm dies.
- ☆ Sporophores of the fungus, *Ganoderma lucidum*, appear at the base of the affected trunk in palms prior to wilting or just after the death of the palm

Integrated Disease management of Ganoderma wilt

- ☆ The diseased part of the garden should be isolated from healthy area by digging isolation trench (1 m. deep and 0.5 m. width).
- ☆ Removal and burning of diseased and dead palms along with roots.
- ☆ The pit for replanting should be filled up with a mixture of soil and farmyard manure in equal quantities along with 50 g of *Trichoderma viride* talc powder + 1 kg neem cake.
- ☆ Seedlings for new plantations should be raised in disease free fields.
- ☆ Injury or damage to roots and pruning or cutting of roots should be avoided to prevent infection through injured roots.

- ☆ Since the disease is more severe in light soils with poor water holding capacity, raising and ploughing in situ of green manure crops like Sunnhemp and Sesbania is advised to increase soil organic matter and antagonistic microflora.
- ☆ Basin / drip system of irrigation to individual palms should be adopted.
- ☆ Application of Talc formulation of *Trichoderma viride* (50 g) in combination with 5 kg neem cake/palm/year should be applied.

Bud Rot (*Phytophthora palmivora*)

Bud rot is a limiting disease that affects most oil palm crops destroying thousands of hectares in Latin America. Bud rot (BR) is caused by the oomycete *Phytophthora palmivora* (Butler). Integrated disease management (IDM) technology has been used to control the disease, which slows down the progress of the disease, allowing palm recovery.

Early detection for Bud rot (BR) management is one but not the only disease management strategy to ensure crop sustainability. It must be integrated with the best agronomic practices, chemical and biological control of the causal agent, and in the long-term, cultivars genetically resistant to the disease.

Symptoms

- ☆ Palms of all age groups are susceptible to bud rot disease
- ☆ The incidence was found to be high in palms below the age of 20 years
- ☆ Withering of spear leaf
- ☆ Rotting and toppling of spindle
- ☆ Infection on nut and nut fall
- ☆ Disease spread from windblown rains

Integrated management strategies evolved for the Management of bud rot disease of oil palm

- ☆ The garden should be kept clean.
- ☆ The trees died due to bud rot should be removed and burnt to avoid further spread.
- ☆ Application of recommended doses of potash fertilizers should be taken up
- ☆ Improve drainage conditions of the field
- ☆ In the initial stages of bud rot, the rotten parts should be removed and destroyed. After thorough cleaning, Bordeaux paste must be applied.
- ☆ The bunches with fruit rot symptoms should be cut down & destroyed. Spray with Bordeaux mixture (1%) or copper oxy chloride (0.3%) should be given.

- ☆ Application of talc formulation of bio control agent, *Pseudomonas fluorescens* according to the age of the palm. 5g, 10g, 75g, 100g, 150g and 200g of talc formulation of *Pseudomonas fluorescens* should be applied in the crown region of seedling aged below one year, one year, 2, 3, 4 and 5 year and above old palms respectively.

Stem Bleeding (*Thielaviopsis paradoxa*)

- ☆ High water table in the soil, higher acidic and alkaline conditions favor the disease.
- ☆ The diseased trees show extensive yellowish brown to black bleeding patches from natural cracks or places of damage due to implements which become infected.
- ☆ The bleeding is commonly seen about more than one foot above the ground level and never seen on the trunk at the ground level.
- ☆ On chiseling of such bleeding patches on young trees of 10-15 years age, yellowish liquid with a fermented smell flows out.
- ☆ Formation of cavities on the trunk is observed in the infected palms.
- ☆ Severe diseased conditions lead to reduction in nut yield and death of the younger palm.

Management of Stem bleeding disease of oil palm

- ☆ Injury or damage to the stem, especially while tractor ploughing should be avoided.
- ☆ Smearing of talc powder paste of *Trichoderma viride* / *T.harzianum* / *T.hamatum* on the bleeding patches on the stem (paste can be prepared by adding 5 ml of water to 10 g of talc powder).
- ☆ Soil application of 50 g of talc powder formulation of *Trichoderma viride*/ *T. harzianum* / *T. hamatum* in combination with 5 kg neem cake / palm /year.

Grey Leaf Blight (*Pestalotia palmarum*)

Drained soils, soils with potash deficiency, continuous rainy weather for 4-5 days and strong winds favour the disease.

Symptoms

- ☆ Initially symptoms develop only on the outer whorl of leaves, especially in older leaves.
- ☆ Minute yellow spots surrounded by a greyish margin appear on the leaflets. Gradually, the centre of the spots turns to greyish white with dark brown margins with a yellow halo.
- ☆ Many spots coalesce into irregular grey necrotic patches.

- ☆ Complete drying and shrivelling of the leaf blade occur giving a blighted or burnt appearance.
- ☆ Large number of globose or ovoid black acervuli appears on the upper surface of leaves.

Management of Grey leaf blight disease of oil palm

- ☆ Remove and burn the infected, fallen leaves periodically.
- ☆ Apply heavy doses of potash.
- ☆ Improve the drainage conditions of the soil.

Bacterial Bud Rot (*Erwinia* spp.)

Symptoms

Parts of spear leaf petiole or rachi turning brown; discoloration may be associated with a wet rot; spear leaf may be wilted and/or chlorotic; leaves may be collapsing and hanging from the crown; infection of the bud results in buds becoming rotten and putrid, leading to death of the palm. Disease occurs in oil palm in Colombia, Costa Rica, Democratic republic of Congo, Ecuador, Nicarauga, Nigeria, Panama and Southeast Asia.

Management

Plant oil palm varieties with resistance to the bacteria; rotting tissue on spear leaves should be removed to prevent bacteria spreading to buds.

Oil Palm Wilt (*Fusarium oxysporum*)

Symptoms

Symptoms of the disease vary with age of host; disease can affect seedlings and mature trees; seedlings exhibit retarded growth, reduced leaf size, chlorosis of older leaves and tip necrosis; field palms may exhibit a bright yellow chlorosis of leaves in the mid-canopy which starts at the tip pf the pinnae and moves towards petioles before affecting adjacent fronds and spreading to older leaves in the canopy; in older palms, lower leaves wilt and dry out and fronds break close to the base of the trunk; new fronds are chlorotic and stunted; the palm shows decline on one side and develops symptoms in the lower canopy; infection spreads rapidly upwards and infects the bud, killing the palm. Fungus infests palms through the root system.

Management

International quarantine procedures have limited the spread of the disease between major palm oil producing countries; dead or dying trees should be felled and burned to prevent spread in plantations; if palms are replanted then new palm should be planted a distance of 3.9 m from infested stump.

Integrated Pest and Disease Management (IPDM) Practices for Oil Palm

Cultural practices

To eliminate the insect pests and diseases or stop them from causing financial harm is one way cultural pest management systems incorporate routine farm tasks. Here is a list of the numerous cultural traditions.

- ☆ Methods used to get nursery or main fields free of insect pests and diseases infestation include removing plant debris, pruning bunds, treating soil, and deep midsummer ploughing (which kills a number of pests).
- ☆ Vitamin deficiency in the soil is tested to determine which fertilizers should be used.
- ☆ Choosing clean, authorized seeds and treating them with bio fungicides or bio-insecticides prior to sowing to control seed-borne disease.
- ☆ Proper plant spacing, which increases plant health and reduces pest risk.
- ☆ It is important to encourage the use of bio fertilizers, FYM, and effective fertilizers.
- ☆ Appropriate water management since excessive wetness in the soil over an extended period of time encourages the growth of pests, particularly those that spread diseases through the soil.
- ☆ Effective weed control as it is a well-known undisputed fact that almost all weeds not only compete with crops for micronutrients but also house a variety of pests.
- ☆ Placing area yellow pan sticky traps for aphids and white flies a few inches above the ground.
- ☆ Synchronized sowing- Here, a network strategy is required to sow the plant simultaneously over a very big area so that pests will not be able to acquire distinctively staged vegetation suitable for increasing their population. If there are too many insect pests and diseases, the procedure is frequently correctly carried out throughout the entire area.
- ☆ Vegetation that is enticingly growing along field edges. It goes without saying that a certain pest species favors some plants over others, and these plants are known to attract that insect pests and diseases. The insect pests and diseases are rise as a result of the vegetation that is grown around the fields' edges; these populations can be reduced either by employing pesticides or by promoting the growth of their natural enemies for herbal management.
- ☆ Intercropping or several crops whenever possible. Every pest species does not appear to be attracted to any particular vegetation, and since desirable

vegetation serves as repellents to keep insect pests and diseases species away from it, the incidence of insect pests and diseases is reduced.

- ☆ Harvesting as close to the stage of the floor. This is because insect pests and diseases with favorable developmental stages remain on plant components, serving as the primary inoculum for the following crop season. Therefore, cutting down on pests in the following season by picking plants at the floor stage.
- ☆ To protect nursery plants from soil-borne diseases, nursery plants are sprayed with copper fungicide or bio pesticide solutions before planting.
- ☆ Throw away crowded, dead, damaged, or diseased branches and destroy them when pruning fruit trees. Avoid piling them inside orchards because this could serve as a source of pest infestation.
- ☆ To shield plants from pest and disease attacks, apply Bordeaux paste or paint to big pruning wounds.
- ☆ Bee hives and floral arrangements designed to attract pollinators promote increased pollination and the subsequent fruit set.
- ☆ The choice of high-yielding pest resistant/tolerant varieties.

Mechanical practices

It is feasible to remove and destroy hundreds of insect pests eggs, larvae, pupae and adults as well as disease-related components of plant life from any location.

- ☆ Putting hundreds of parasitized eggs within bamboo cages with hen perchers installed inside the sphere to preserve natural enemies and whenever feasible, keep insect pests and diseases species away.
- ☆ The use of light traps and the eradication of captured insects.
- ☆ Installing the necessary hen scarer inside the sphere and adding hen perches inside the sphere to allow birds to perch there while navigating bugs and their juvenile stages, such as eggs, larvae and pupae.
- ☆ The use of pheromones to create new areas and stop mating and pheromone traps to monitor and control the population of insect pests and diseases by mass capturing using.

Regulatory practices

In this procedure, governmental regulations are put into effect that prevent seeds and plant materials that have been infected from entering the country or moving from one region to another. These are referred to as quarantine techniques and come in two flavours: domestic and international.

Biological practices

The most important aspect of IPM is biological control of diseases and insect pests and diseases through biological methods. Bio-control, in a more general meaning, is the use of living things to control other living things (pests). In other words, the deliberate use of parasitoids, predators, and diseases to control pest populations at a level below those resulting in economic loss, either by introducing a new bio agent into the environment of pests or by increasing the effectiveness of those already present in the field.

Conclusion

The sustainability of oil palm cultivation is the need of the hour for farmers as well as processors and finally the enrichment of the economy. To minimize losses, it is crucial to control the insect pests and diseases population below economic threshold levels. In order to achieve this, integrated management techniques that improve the sustainability of the oil palm industry as a whole must be established employing the most recent low-cost technologies.

References

Dhileepan, K. 1990. Trials on the protection of Oil palm fruit bunch from bird damage in India. *The Planter* **66**: 171-177.

Dhileepan. K.1992. Insect Pests of oil palm *(Elaeis guineensis)* in India. *The Planter,* Kuala Lumpur, **68**: 183-191.

Kalidas, P. 2019. The inference of the impact of rugose spiraling whitefly on oil palm fresh fruit bunch yield in India. *Planter,* **95**(1115), 83-89.

Murphy, D.J., Goggin, K and Paterson, R.R.M. 2021. Oil palm in the 2020s and beyond: Challenges and solutions. *CABI Agric. Biosci,* 2: 39.

Nair, C.P.R. and Daniel, M 1982. Pests. In: *The Arecanut Palm, A monograph* (Eds. Bavappa, K. V.A., Nair, M.K. and Premkumar, T.) Central Plantation Crops Research Institute, Kasaragod, India.

Chapter-6

BIO MANAGEMENT OF COCOA DISEASES AND INSECT PESTS

B. Neeraja[1], N. Emmanuel[2], and A. Kireeti[1]

[1]Dr YSRHU-Horticultural Research Station

Dr. YSR Horticultural University

Ambajipeta, Dr. B. R. Ambedkar Konaseema District, A.P.

[2]Dr YSRHU-College of Horticulture

Dr. YSR Horticultural University

Venkataramannagudem, West Godavari District, A.P.

Introduction

The cocoa tree (*Theobroma cocoa* L., Malvaceae) grows in the understory. It produces fruits (pods) containing seeds (beans) that are used in the confectionery industry, particularly in the production of chocolate. Cocoa beans, the seeds of the *Theobromae* cocoa tree (a native of South America's Amazon region), are widely consumed around the world. It is a significant Neotropical, perennial crop that grows between 20° north and 20° south of the equator (World Cocoa Foundation Report, 2014). Cocoa production is critical for 5.5 million smallholder cocoa farmers and 14 million rural workers (Fountain and Hutz-Adams, 2015).

Cocoa, the food of the Gods, has served humanity for many decades. The use of cocoa-based products is becoming more popular. Indeed, the high polyphenol content of cocoa, combined with its widespread presence in many food items, makes this food of particular nutritional and health interest. Cocoa farming provides an income for approximately 40 to 50 million people in India. Cocoa is primarily grown in Kerala, Karnataka, Andhra Pradesh, and Tamil Nadu as a tropical crop with significant development potential. Kerala is the most populous of the aforementioned states.

The increasing popularity of chocolate and the drink cocoa is self-evident, as evidenced by the many "chocolate lovers" who look forward to special occasions and national holidays when chocolate is traditionally given as a gift to loved ones and friends. The demand for chocolate is increasing, and the question now is whether the world supply of a product derived solely from a tropical, rainforest-dwelling tree will be sufficient to meet the demand. Several important fungal diseases and pests have gained prominence in the last two decades and pose a serious threat to the supply of chocolate.

Bio-control of Insect Pests and Diseases in India

Biological control was discovered through trial and error and was used in agriculture long before the term was coined (Baker and Cook, 1974). For example, the ancient practice of not growing the same crop species in the same field more than every second or third year, or even longer, is an example. Because of the predatory, competitive, and other antagonistic effects imposed by the associated microflora and fauna, crop rotation allows time for the pest or pathogen population in soil to decrease below some economic threshold. In other words, crop rotation gives the natural soil microbiota time to sanitize the soil, which is especially important for more specialised plant parasites and insect pests that rely heavily on their host crop to survive.

The introduction of the vadalia beetle from Australia to California in 1888 to control the cottony cushion scale of citrus brought in the modern era of biological control, which involves the deliberate transfer and introduction of natural enemies of insect pests. In 1914, the German plant pathologist C. F. von Tubuef published a somewhat speculative article titled "Biologische Bekampfung von Pilzkrankheiten der Pflanzen." Baker (1987) claims that this is the first time the term "Biologische Bekampfung" or "biological control" appears in the scientific literature.

Biological control, according to DeBach (1964), is "the action of parasites, predators, or pathogens in maintaining another organism's population density at a longer average than would occur in their absence." This definition includes some highly successful biological controls of insect pests with natural enemies, but it excludes some other highly successful controls accepted as examples of biological control in other disciplines. Citrus tristeza virus, for example, is controlled in Brazil by inoculating citrus trees with a mild virus, which then protects the trees from more severe strains (Costa and Muller, 1980). H. H. McKinney demonstrated the potential of "cross protection" for biological control of plant viruses in 1929. Plant pathologists refer to cross protection as a method of controlling plant viruses.

Need for Biological Control

To meet the needs of the growing population, food grain production should increase to 250 million tonnes by 2020. Growers frequently rely on chemical fertilisers and pesticides in addition to good agronomic and horticultural practises. However, environmental pollution caused by excessive use and misuse of agrochemicals, as well as the propaganda used by some pesticide opponents, has resulted in significant shifts in people's attitudes towards pesticide use in agriculture.

As the proportion of pests and diseases increased, so did the use of toxic chemicals for their management. The number of pesticide and fungicide-resistant species is growing. In recent years, following the signing of the World Trade Organization's General Agreement on Trade and Tariffs, more emphasis has been placed on the use of ecofriendly pesticides for crop production due to their less toxic nature, low levels of disease resistance, and low residue problems. Biological controls, on the other hand, should be combined with other control measures because different methods are effective at different times and locations under different conditions.

Pest and disease management is a critical component of cocoa health management. In India, more than 50 insect pests have been identified as infesting cocoa (Mariamma Daniel, 1994). Pest and disease outbreaks are common during the summer and post-monsoon seasons. In Andhra Pradesh, cocoa is grown as an intercrop in the coconut ecosystem on 13720 hectares. The majority of the land is in the erstwhile East and West Godavari districts of Andhra Pradesh, covering 9512 and 2520 hectares, respectively.

Bio-control of Cocoa Diseases

Phytophthora pod rot / Black pod rot disease: *Phytophthora palmivora*

Cocoa black pod rot is caused by a pathogen in the genus *Phytophthora*, which means as "plant destroyer." This is the genus that caused the Irish potato famine of 1845-1852. These pathogens were originally classified as fungi, but they have since been reclassified as Stramenopila. There are over 80 *Phytophthora* species that cause plant diseases, with several, including *P. palmivora*, *P. megakarya*, *P. citrophora*, and *P. capsici*, being responsible for cocoa black pod rot. *Phytophthora* spp. is responsible for 20 to 30% of the total cocoa crop loss each year, with some plantations losing up to 90% of their pods due to the disease (Acebo-Guerrero, 2012).

P. palmivora has the ability to attack pods at any stage of development (Thorold 1975). Sporangia or zoospores that germinate and penetrate the pod surface may cause infection. Disease symptoms may appear 3-4 days after infection, and sporangia production may begin on the night following the second day of visible symptoms (Spence 1959). Adegbola (1981) described a typical black pod in addition to the 'seaweed-like' odour that characterises pod rot caused by *P. palmivora* (Sreenivasan and Quesnel, 1977). The disease's most common symptom is fruit rot, also known as black pod or *Phytophthora* pod rot. However, stems and trunks can also be affected, with cankers appearing on the bark and flower cushions disappearing. Infections have been noticed on leaves and roots as well, although less frequently and with a much lower frequency.

Fruit with brownish spots. When *P. palmivora* infects cocoa, the pathogen penetrates the waxy cuticle and attacks the epidermis. At the site of infection, a small brownish spot appears. Such infections can start at the stem or blossom ends of fruits and spread to the fruit. Infection spreads quickly across the outer surface, eventually covering the entire pod in a matter of days. Infected areas turn brown to black and if conditions are favourable, clusters of white sporangia appear on the pod's outer surfaces. As visible symptoms progress, the pathogen moves deeper into the pod, infecting and destroying the beans. Infected beans quickly deteriorate and rot, rendering the pod useless. The pods then dehydrate and mummify.

Under favourable environmental conditions such as those required for optimal cocoa tree growth, the mass of sporangia on the fruit surface may serve as a source of primary inoculum for disease spread (Adegbola 1981). Thorold (1953) and Blencowe and Wharton (1961) conducted field surveys that revealed that yield could play a significant role in the development of black pod disease epidemics. Their findings indicated that black pod loss increases as yield increases.

Distal infection, on the other hand was most common in young pods due to the retention of moisture at the distal end. Most early infections were either proximal or distal but as inoculum accumulated in the field, lateral infection could occur. Young pod (cherelle) infection is frequently confused with cherelle wilt. According to Opeke and Gorenz (1974), the latter is distinguished from black pod by the presence of a preliminary yellowing followed by slow broadening from the tip.

Biological control

Extensive research has been conducted to discover and apply conventional inundative biological control agents against *Phytophthora* diseases of cocoa. Despite numerous reports of antagonistic and mycoparasitic fungi inhibiting *Phytophthora* growth in vitro, no commercial products have been released or widely adopted by cocoa farmers. *Phytophthora*'s short life cycle, phenomenal reproductive capacity, complex disease cycle, and zoospore motility produce explosive epiphytotics in cocoa, making inundative biological control agents ineffective.

More believe can be gained from the discovery and development of antagonistic endophytes. In the natural cocoa forest ecosystem, endophytic fungi are naturally transmitted from mature trees to seedlings; however, they eventually disappear from plantations. Recent research indicates that antagonistic endophytes reintroduced into cocoa survive and protect the tree from *Phytophthora*. Endophytes have the potential to play a significant role in integrated disease and pest management programmes. The use of mulches and composts improves soil health and microbial activity while suppressing *Phytophthora* and these methods are also important components of integrated management programmes.

Stem Canker disease in cocoa - *Phytophthora palmivora*

It affects stem, flower cushions, jorquette and fan branches. In early stage of disease development it is difficult to detect the symptoms and so close observation is required. Stem canker disease symptoms appears after monsoon. Shade and more humidity favours the disease.

Symptoms

- ☆ Dark brown, round to oval shaped discolored area are formed on the bark.
- ☆ Oozing of reddish brown liquid from the infected area.
- ☆ Girdling of infected stem.
- ☆ At the collar region canker lesion will be bigger and it spreads fast and infects the below ground root system and affects water and nutrient uptake from plants.
- ☆ Yellowing and wilting symptoms followed by death of plant.

Management

- ☆ Phytosanitation practices like removal and destruction of black pod infected pods and fallen leaves.
- ☆ If mulching is practiced, avoid trunk covering or mulch 1 feet away from the trunk.
- ☆ Providing good drainage facility in the garden.
- ☆ Proper pruning twice in a year.
- ☆ Wound treatment and soil application with Trichoderma coir pith cake, helps in long term management of the disease.

Charcoal pod rot (*Lasiodiplodia theobromae*)

Lasiodiplodia theobromae is a common widespread pathogen of tropical woody trees, causing shoot blight and dieback of trees and shrubs and blue stain in timber (Mohali *et al.*, 2005). *Lasiodiplodia theobromae* was identified based on morphological characters as described by Punithalingam (1976). These included A rapidly spreading dark green or black colony was present as were conidia that were dark brown, striate, ellipsoidal, uniseptate and produced in ascostromatic pycnidia on potato dextrose agar (PDA).

Symptoms

- ☆ Occurs throughout the year, becomes severe during summer months.
- ☆ Pods of all ages are susceptible.
- ☆ Infection appears as dark brown to black coloured spot. The affected spots turn black and remain on the tree as mummified fruit.
- ☆ The internal tissues are rotten and the affected beans turn black.
- ☆ Spores appear in masses forming soot.
- ☆ Infection takes place through wounds.

Insect pests of cocoa

In India cocoa is known to be attacked by about 50 insect and non insect pests comprising mealy bugs, tea mosquito bug, stem borer, aphids, stem girdler, leaf eating caterpillars and leaf eating beetles etc. among the insect pests . Recently, Emmanuel *et al.,* (2011) have reported the occurrence of chaffer beetles, *Adoretus versutus* and *Apogonia blanchardi* and bagworms, *Pteroma plagiophelps* and *Clania* sp on cocoa crop.

Table 1: List of cocoa insect pests

Common name	Scientific name
Chaffer beetles	
Indian Rose beetle	*Adoretus versutus*
Black leaf chaffer beetle	*Apogonia blanchradii*
Bag worms	*Pteroma plagiophelps* & *Clania* sp.
Leaf eating caterpillars	
Hairy caterpillars	*Euproctis subnotata, Euproctis fraternal*
Tussock caterpillar	*Dasychira mendosa*
Indian Gypsy moth	*Lymantria obfuscate*
Slug caterpillar	*Parasa lepida*
Looper caterpillar	*Thalassodes* spp
Tobacco caterpillar	*Spodopetra litura*
Bark eating caterpillar	*Indarbella Obliquifasciata*
Sucking pests	
Mealy bugs	*Planococcus lilacinus, Planococcus citri, Paracoccus marginatus, Rastrococcus iceryoides*
Aphids	*Toxoptera aurantii* and *Aphis gossypii*
Non insect pests	
Rats	*Rattus rattus*
Squirrels	*Funambulus trisriatus* and *F. palmarum*

Chaffer Beetles

☆ **Brown leaf chaffer beetle/Indian Rose beetle [*Adoretus versutus*]**: *A. versutus* can cause heavy damage in nurseries and young cocoa plantations. Seedlings are very vulnerable to attack because they can be defoliated rapidly. Rose beetles attack 3 to 4 month old seedlings in nurseries and young plants after planting out (Beaudoin, 1992).

A. versutus adults feed by perforating the leaflets starting from the middle and without destroying the ribs. The leaflets are eaten away in very small but numerous patches, giving a skeletal appearance to the leaflet. The attacks are more numerous at the apex of the leaflets than at the base. Besides this characteristic feeding behaviour, the adult rose beetles make depressions in the border of the areas eaten, which is typical of *Adoretus* spp. and distinguishes them from the damage caused by other foliage pests. *A. versutus* feed in the early hours of the night.

☆ **Black leaf chaffer beetle [*Apogonia Blanchardi*]:** Black leaf chaffer beetle were observed feeding on the cocoa foliage during night times confirming to be nocturnal. Adults feed on the cocoa leaf from the peripheral region. Adult is deep black species without metallic luster. Length is 9.5 to 11 mm, ovate, glabrous, shining black, the apex of the elytra and the two basal ventral segments, however, opaque; the antennae and palpi pale ferruginous ; the under surface and legs sprinkled with pale coloured setae.

Management of Chaffer beetles

Spraying of Azadirachtin 10000 ppm @ 1ml per liter of water.

Bag worms [*Pteroma plagiophelps* and *Clania* sp]

The self enclosing bags of *P. plagiophelps* is 1.8 to 2 cm long and 5 to 6mm wide. The caterpillar builds silky bag in narrow cone shape with plant materials live in this mobile case and rest by sticking top opening of the case to the cocoa leaves and hang their bag vertically. Whereas the self enclosing bags of *Clania sp* is 3.5 to 4.4 cm long and 8 to 9mm wide .This species uses sticks of similar size and there are about 9 – 13 sticks arranged parallel around the silk case except one or two of their sticks used are much longer than the other. The caterpillar lives inside the case and feeds on the cocoa foliage. Both the *P. plagiophelps and Clania sp* feeds the cocoa leaves from the central leaf lamina in a circular to irregular holes.

Leaf eating caterpillars

☆ **Tussock caterpillar [*Dasychira mendosa*]:** The larvae of this moth feed on foliage of many plants including cocoa. Therefore, due to this large host range, the breeding continues throughout the year during which there are probably 5 or 6 generations. The moth, which has pale yellow, hind wings and forewings are irregularly patterned with various shades of brown, lays large masses of eggs. The feeding by larvae results in defoliation and the larval period lasts for 21-28 days. The fully-grown hairy larva has a reddish head; the body greyish or yellowish, is striped with red and with long dense dorsal tufts of whitish hairs. They pupate in loose cocoons made of silk and hairs and the pupal period lasts for 11-12 days.

☆ ***Euproctis subnotata*:** Caterpillars are small with dark brown in colour with yellow bands and minute hairs. Pupation takes place in soil and adults are brownish in colour. Forewings are fibrous and hindwings are yellowish in colour. Laboratory studies with *Euproctis subnotata*, reared on cocoa leaves, showed that the average incubation period was 6.7 days, larval period 23.57 days, pupal period 11 days, total life cycle 35.77 days and. Male adult longevity was 6 days, while the female adult longevity was

8 days. Females laid an average of 155 eggs and the larvae underwent 6 instars.

☆ ***Euproctis fraterna***: Reddish brown hairy caterpillar, head red in colour surrounded by white hairs. Abdomen has tufts of hairs all over the body and a long pre anal tuft. While adult is yellow moth with pale transverse lines and black spots on the forewing. The development of *Euproctis fraterna* was studied on cocoa leaves .The mean incubation period was 6.93 days. The larvae passed through 6 instars and completed development in 23.25 days. The pupal stage occupied 11 days. The total life cycle was completed in 41.18 days. Male adult longevity was 6 days, while the female adult longevity was 8 days. Females laid an average of 146 eggs.

☆ ***Lymantria obfuscata***: In adults of gypsy moth, sexual dimorphism is well distinct. Female moths larger in size and dirty creamish whereas males smaller than females and have brownish bodies. In males wings are well developed. Front and hind wings dissimilar in venation and in shape. Larval body covered with tuft of hairs, bearing a pair of small dots on each segment running all along the back. Laboratory studies with *Lymantria obfuscata*, reared on Cocoa leaves, showed that the average incubation period was 8.7 days, larval period 26.57 days, pupal period 11 days, total life cycle 36.77 days and male adult longevity was 5 days, while the female adult longevity was 7 days. Females laid an average of 158 eggs and the larvae underwent 6 instars

☆ ***Euproctis scintillans***: The larvae of this species feed voraciously on leaves. Initially the larvae scrape the green matter resulting in skeletonization. Later the larvae move into the other parts of the plant and defoliate. This hairy caterpillar larval stage lasts for 13-29 days and the pupal period lasts for 9-20 days. The total life cycle lasts for 6-7 weeks.

Slug caterpillar [***Parasa lepida***]: Adult moth has green wings with prominent dark patch at the base of each forewing. Larva is characterized by greenish body with white lines and four rows of spiny scoli tipped red or black, which cause irritation and pain. Eggs of *Parasa lepida* are laid on leaves which hatch in 3-5 days. Young larvae feed on the leaf epidermis and as they develop, chew up leaves. There are 7 larval instars (often 8 instars for females) which are completed in 35-42 days. Pupation occurs in cocoons often attached to stem or bark and the pupal stage lasts 21-24 days.

Looper caterpillar [*Thalassodes* spp]: The adult is a stout moth exhibiting sexual dimorphism. The female moth is fairly bigger (body length 29-31 mm) than adult male (body length 25 mm). Male moths have a black patch with a small white spot in the centre of fore-wing and two angulated transverse lines on either side of the patch, but there are no black patches present in the fore-wings of the females, except short wavy lines running across. The abdomen of the male is long and slender; whereas it is stout with tuft of anal hairs in female. This looper larvae are found feeding on developing leaves. The larval period lasts for 17-18 days. The larva possessing the color of new shoots and assuming a characteristic pose on the twig is often mistaken for a leaf petiole. The pupal period lasts for 7-8 days. The total life cycle was completed in 33.18 days.

Management of Leaf eating caterpillars

- ☆ Spraying of Azadirachtin 10000ppm @ 1ml/L or Azadiractin 1500ppm @ 5ml/L.

Bark eating caterpillar [*Indarbela obliquifasciata*]

It is a Polyphagous pest reported on Citrus, mango, guava, jaman, loquat, mulberry, pomegranate, ber, litchi and aonla. Adult is stout yellowish –brown moth with brown wavy markings on the forewings. Males are smaller than the females. Young trees may succumb to the attack. Caterpillars bore into the trunk or junction of branches. Presence of gallery made out of silk and frass. Caterpillars remain hidden in the tunnel during day time come out at night and feed on the bark. Older trees and the trees in orchards that are not well maintained are more prone to these pests. Peak activity period is September-October. Eggs are laid under loose bark or in cracks and crevices in clusters of 15-25 from April to June. They hatch in 8-11 days. Larval duration of 9-10 months. Pupal period is 3 to 4 weeks.

Management

- ☆ Orchard should be kept clean and overcrowding of trees should be avoided
- ☆ Remove and destroy dead and severely affected branches of the tree.
- ☆ Remove alternate host, silk cotton and other hosts.
- ☆ Swab Coal tar + Kerosene @ 1:2 on the basal portion of the trunk up to 3 feet height.
- ☆ Scraping the loose bark to prevent oviposition by adult beetles.

Mealybugs: (*Planococcus lilacinus, Planococcus citri, Paracoccus marginatus and Rastrococcus iceryoides*)

It colonizes on the tender parts of the plant such as growing tips of the shoots, the terminal buds, the flower cushions, the young cherelles and mature pods. Feeding of mealy bugs induces cherelle wilt.

Aphids [*Toxoptera aurantii* and *Aphis gossypii*]: They colonize on the underside of tender leaves, succulent stem, flower buds and small cherelles. Heavy infestation may occur during hot summer and after rainy season which brings about premature shedding of flowers and curling of leaves.

Management of sucking pests

- ☆ Spraying of Azadirachtin 10000ppm @ 1ml/L or Azadirachtin 1500ppm@ 5ml/L.

Rodent pests in cocoa

Rodents are serious non insect pests in fields and in human dwellings, both in the countryside and in cities. Rodents cause direct damage to various crops/commodities by gnawing and feeding and indirect damage by spoilage, contamination and hoarding during on-farm and post harvest stages. Rodents are characterized with quadri-pedal body covered with hairs and presence of one pair of chisels shaped ever growing incisor teeth in each jaw while canine teeth are absent leaving a wide gap called diastema.

Plantation crops such as coconut, cocoa and oil palm often face rodent damage. Knowledge of the characteristics, extent of damage and vulnerable period for rodent attack in coconut and cocoa is important for appropriate planning of management strategies. Rats cause damage near the peduncle of the cocoa pods while squirrels make damage to the cocoa pods in the centre. The intensity of damage to cocoa by rats and squirrels is always at higher side.Rodents- Rats (*Rattus rattus*) and Squirrels (*Funambulus tristriatus* and *F. palmarum*) these vertebrate pests feed on the mucilage covering of the beans and cause serious damage to the pods.

Rat damage

Symptoms

- ☆ The rats usually gnaw the pods near the stalk portion.
- ☆ The rats are known to damage the mature as well as immature cocoa pods .
- ☆ Management: rat traps can also be used with fried coconut pieces.

Squirrel damage

Symptoms

- ☆ Squirrels gnaw the pods in the centre.
- ☆ Squirrels damage only the mature ones.

Management:

- ☆ Squirrels are best controlled by trapping with wooden or wire mesh single catch 'live' trap with ripe coconut kernel as the bait.
- ☆ The success is more if trapping is carried out during the lean periods of the crop (October-November) and when the alternate foods such as paddy, cashew apples, mangoes and jackfruits are not available.
- ☆ Timely harvest of the pods as well as maintaining proper plant density will help in increasing the efficiency of poison baiting as well as trapping.

For growth of agricultural production has led several new challenges, making further growth possible only if these challenges are met appropriately and timely. Increase in crop production from the modern farming techniques reaching a plateau is the most of the countries including India and the environmental problems due to excessive use of chemical fertilizers and pesticides becoming a matter of concern. So, the biological control can be alternate system, which may play an important role in achieving the goal of agriculture.

References

Acebo-Guerrero, Y., Hernández-Rodríguez, A., Heydrich-Pérez, M., El Jaziri, M. and Hernández-Lauzardo, A.N. 2012. Management of black pod rot in cacao (*Theobroma cacao* L.): a review. *Fruits, 67*(1), pp.41-48.

Adegbola, M.O.K. 1981. Cocoa diseases of West Africa. In *7. International Cocoa Research Conference. Proceedings, Douala, Cameroon, 4 12 Nov 1979.* (pp. 243-250). Cocoa Producers' Alliance.

Baker, K.F. and Cook, R.J. 1974. *Biological control of plant pathogens.* WH Freeman and Company.

Baker, K.F. 1987. Evolving concepts of biological control of plant pathogens. *Annual review of phytopathology, 25*(1), pp.67-85.

Beaudoin, L. 1992. Feasibility study into the biological control of the rose beetle *Adoretus versutus* Harold within South Pacific.

Blencowe, J.W. and Wharton, A.L. 1961. Black-pod disease in Ghana: incidence of the disease in relation to levels of productivity. *Black-pod disease in Ghana: incidence of the disease in relation to levels of productivity.*

Blencowe, J.W. and Wharton, A.L. 1961. Black-pod disease in Ghana: incidence of the disease in relation to levels of productivity. *Black-pod disease in Ghana: incidence of the disease in relation to levels of productivity.*

Costa, A. S., and G. W. Muller. 1980. Tristeza control by cross protection: A U.S.-Brazil cooperative success. Plant Disease 64:538-541.

DeBach, P. 1964. Biological control of insect pests and weeds. *Biological control of insect pests and weeds.*

Emmanuel, N., Chalapathi Rao, N.B.V. and Gautam, B. 2011. New record of defoliator pests of cocoa in Godavari districts of Andhra Pradesh and their management.

Fountain, A. and Hütz-Adams, F. 2015. Cocoa Barometer (USA edition)https://www.cocoanet.eu/uploads/media/ACVV_FGTB_Horval_October_2014_WCC2014.

Mariamma Daniel. 1994. Pests of Cocoa. In - Advances in Horticurture Vol. 10. Plantation and Spices Crops. part 2 (eds. Chadha, K.L and. Rethinam, P). Malhotra publishing House, New Delhi. pp 743-758.

McKinney, H. 1929. Mosaic diseases in the Canary Islands, West Africa and Gibraltar. *Journal of Agricultural Research.*, **39**: 577-578.

Mohali, S., Burgess, T. I. and Wingfield, M. J. 2005. Diversity and host association of the tropical tree endophyte *Lasiodiplodia theobromae* revealed using simple sequence repeat markers. *Forest Pathology* **35**: 385–96.

Opeke, L.K. and Gorenz, A.M. 1974. Phytophthora pod rot: symptoms and economic importance. *Phytophthora Disease of Cocoa. PH Gregory, ed.*

Punithalingam, E. 1976. *Botryodiplodia theobromae.* CMI Descriptions of Pathogenic Fungi and Bacteria No. 519. Wallingford, UK: CAB International.

Spence, J. A. 1959. The problem of *P. palmivora* on the cocoa plant. Journal of Agricultural Science (Trinidad), **59**: 97–111.

Sreenivasan, T.N. and Quesnel, V.C. 1977. Field differentiation of black pod from other cacao pod diseases. *Tropical Agriculture.*

Thorold, C.A. 1953. The control of black pod disease of cocoa in the western region of Nigeria. In *Proc. West African International Cocoa Research Conference, 12th-16th December 1953* (pp. 108-115).

Thorold, C.A. 1975. *Diseases of cocoa.* Clarendon Press.

Indian Rose beetle (*Adoretus versutus*)

Black leaf chaffer beetle (*Apogonia blanchradii*)

Bag worm [*Pteroma plagiophelps*]

Bag worm [*Pteroma plagiophelps*]

Chaffer beetle damage on cocoa

Bark eating caterpillar (*Indarbela* sp.)

Bark eating caterpillar (*Indarbela* sp.)

Euproctis fraterna

Euproctis subnotata

***Phytophthora* canker**

Pod rot

***Botryodiplodia* pod rot**

Chapter-7

BIO MANAGEMENT OF INSECT PESTS AND DISEASES OF CASHEW

B. Nagendra Reddy[1], V. Anoosha[2] and K.Umamaheswara Rao[1]

[1]Dr YSRHU-Cashew Research Station
Dr. YSR Horticultural University
Bapatla, Bapatla District, A.P.
[2]Dr YSRHU-Horticultural Research Station
Dr. YSR Horticultural University
Ambajipeta, Dr. B. R. Ambedkar Konaseema District, A.P.

Introduction

Cashew *Anacardium occidentale* L. Anacardiaceae) is a very important foreign exchange earning crop of India. India is the largest cashew exporter, with more than 15% of the world's export share. During 2021-22, the country's exports grew by 7% to US$ 452 million from US$ 420 million in 2020-21. The cashew industry has large economic significance as it employs more than 10 lakh people on farms and factories in rural areas. It was originally introduced into India from Brazil by Portuguese travellers during sixteenth century mainly for checking soil erosion on the coast. Apart from India, cashew is grown in Ivory Coast, Brazil, Indonesia, Vietnam, Nigeria, Benin, Guinea Bissau, Mozambique, Philippines etc. In India, it was first introduced in to Goa, from where it has spread along the coastal regions of the peninsula (Pillai *et al.,* 2008). In addition, cashew is also grown in few pockets of Assam, Chhattisgarh, Gujarat, Meghalaya, Nagaland and Tripura. Commercial cultivation of cashew is taken up in eight states of our country mainly in west and eastern coast *viz.,* Andhra Pradesh, Goa, Karnataka, Kerala, Maharashtra, Odisha,

Tamilnadu and West Bengal and some parts of the North-Eastern region (Elakkiya *et al.*, 2017). The cultivation of cashew in India covers a total of 0.7 million hectares of land, which is the highest among cashew growing countries and the annual total production is approximately 0.4 million metric tonnes annually (Haritha and Ushadevi, 2022). Every part of cashew tree is infested by one or the other pest. Cashew is reported to be infested by more than 180 insect pests in India. In which, tea mosquito bug and cashew stem and root borer are the two important pests. Besides, there are also various insect pests that attack cashew at different stages of growth. However, depending on the climate, location and age of the plantation, each geographic region has its own distinctive pest complex (Vanitha and Raviprasad, 2020).

Insect Pests of Cashew

Tea mosquito bug (*Helopeltis antonii*)

Tea mosquito bug is considered as most important and serious pest of cashew in almost all cashew growing countries of the world. In India, this pest is common in most of the cashew growing regions including Kerala, Karnataka, Goa, Maharashtra, Tamil Nadu, Andhra Pradesh, Odisha, Madhya Pradesh. Among the 41 recognized species of *Helopeltis* distributed in the world (Stonedahl, 1991; Stonedahl, *et al.*, 1995). Three species (*H. antonii* Sign., *H. theivora* Wat. and *H. bradyi* Wat.) are recorded from India. It causes more than 30% economic loss by inflorescence blight and immature nut fall.

Symptoms of damage

Both nymphs and adults suck sap from the leaves, young shoots, inflorescences, growing young nuts and apples. Each nymph/adult during its life time damages at least three tender shoots/ panicles. Feeding by the nymphs and adults causes drying up of new flushes, panicles, shriveling of developing nuts and premature drop of immature nuts. Bug feeding on developing apples and nuts causes brown sunken spots. Feeding lesions coalesce and the tender shoots and the entire panicles having tender immature apples and nuts wilt subsequently resulting in burnt-up appearance known as "shoot blight or blossom blight". Severely affected branches may lead to the secondary infection by fungus (*Botrydiplodia theobromae*) causing die back disease and subsequent development of numerous auxiliary buds causes a bunched terminal growth known as 'witches broom'. The growth of trees is seriously retarded and fruit formation of attacking flowering shoots is reduced (Vanitha and Raviprasad, 2020).

Biological control of Tea mosquito bug

Parasitoids

The eggs of the tea mosquito bug are naturally parasitized by *Telenoums cuspis* (Hymenoptera: Platygastridae), *Erythmelus helopeltidis* (Hymenoptera:

Mymaridae), *Ufens* sp. (Trichogrammatidae) and *Gonatocerus* sp. (Mymaridae). Nymphal stage of this pest is parasitized by *Leiophron* sp. (Braconidae).

Predators

Spiders such as *Oxyopes schireta*, Phildippus patch and *Hiyllus* spp. are efficient predators and feed on nymphs and adult mosquito bug. Red ants *Oecophylla smaragdina* should be encouraged in cashew plantations as it will repel the tea mosquito bug.

A wide number of reduviid bugs, praying mantids, mantispid flies, robber flies, pentatomid bugs etc also predate on several insect pests.

Leaf miner: *Acrocercops syngramma*

Symptoms of damage

Young cashew plants are observed to be prone to attack by this pest. Caterpillars cause mining injury both in tender leaves as well as shoots. They mine and feed below the epidermal layer of the tender leaves causing extensive leaf blisters which later dry up causing distortion, browning curling of the leaves and drying out of the damaged portion.

Biological control of Leaf miner

Larval parsitioids *viz., Chrysocharis* sp., *Closterocerus* sp., *Sympiesis* sp. and *Aprostocetus* sp. (Eulophidae: Hymenoptera) were recorded on cashew leaf miner larvae. *Chelonus* sp. (Braconidae: Hymenoptera) also parasitizes the larvae of cashew leaf miner.

Cashew stem and root borer (CSRB)-*Plocaederus ferrugineus, Plocaederus obesus and Batocera rufomaculata*

Symptoms of damage

P. ferrugineus is the commonly occurring species. It is a hidden and important pest of cashew, capable of killing even the grown up trees. Symptoms of damage become visible only at the later stages of infestation. Mango stem borer, *B. rufomaculata* attacks the cashew trees at later stage of CSRB infestation.

Biological control of Cashew stem and root borer

Egg Parasitoid, *Avetianella batocerae* Ferriere (Encyrtidae: Hymenoptera) parasitizes the eggs of CSRB. Entomopathogenic fungi namely *Metarhizium anisopliae* and *Beauveria bassiana* infects Later stages of CSRB grubs. The entomopathogenic nematodes viz., *Steinernema* and *Heterorhabditis* are found to be effective in inducing mortality of grubs.

Shoot tip cater pillar: *Anarsia epotias and Hypotima (Chelaria) haligramma*

Symptoms of Damage

During flushing period of cashew, caterpillars web the tender leaves together and feed within at the early stages. Later on, they bore in to the terminal shoots and tunnel inside up to 2-3 cm. A gummy substance oozes out from the infected tips leads to drying of shoots which results in production of auxiliary shoots.

Biological control of shoot tip caterpillar

Larval parasitoids, *viz., Pristomerus* (Ichneumonidae), *Apanteles* sp. (Braconidae), *Elasmus* sp. (Elasmidae) and *Sympiesis* sp. (Eulophidae) parasitizes the on shoot tip caterpillar.

Apple and nut borer, *Thylacoptila paurosema*

Symptoms of damage

- ☆ In the early stages, young larvae move to the joints of nut and apple scrape the epidermis and then bore into them.
- ☆ In later stages, they bore into tender apples and nuts and feed on them. The borer affected nuts do not develop, become shrivelled and dried up resulting in pre mature fall of nuts and apples.
- ☆ Caterpillar tunnels near the junction of apples and nuts, and the entry holes are plugged with excreta. Damaged fruits can be easily located as they have frass hanging externally at fruit and nut joint.

Biological control of Apple and nut borer

Three larval parasitoids *viz., Panerotoma* sp., (Braconidae) and *Trathala flavorbitalis* (Ichneumonidae) on apple and nut borers.

Leaf and blossom webber, *Lamida (Macalla) moncusalis*

Symptoms of damage

- ☆ Caterpillars feed on the terminal leaves of new shoots and blossoms. Galleries of silken webs reinforced with plant scraps and castings, indicate the presence of caterpillars.
- ☆ Presence of webbing on terminal portions, with clumped appearance, and drying of webbed shoot/ inflorescences.

Biological control of Leaf and blossom webber

- ☆ Elasmids (*Elasmus* sp., *Elasmus johnstonii* F.), Braconids (*Apanteles* sp.), and Tachinids (*Blepharella lateralis*) can cause larval parasitism up to 50 per cent in east coast regions.

- ☆ *Apanteles* sp., *Avga choaspis* (Nixon) (Braconidae) parasitize leaf and blossom webber larva in Kerala.
- ☆ Green lace wing, *Chrysoperla* sp. also predate in this pest.

Leaf folder and Leaf rollers (*Hypatima haligramma* , *Caloptilia tiselaea* , *Dudua aprobola* , *Sylepta derogatta*)

Symptoms of damage

- ☆ Larva of *Hypatima* roll up the leaf margin towards ventral side. Roll is narrow and tight as a result leaf dries up.
- ☆ Larva of *Caloptilia* folds loosely the leaf either from the top or from lateral margin towards the midrib ventrally. The larva remain inside the roll and feed from inside by scraping the green portion of the leaf lamina which later dries up.
- ☆ Larva of *Dudua aprobola* damage tender leaves by making spindle shaped folds. Two to four terminal leaves are folded longitudinally one above the other and fastened with silken threads to form a tight tubular roll at the growing point resulting in delayed panicle emergence.
- ☆ Larvae of *Sylepta auranticollis* during their early stages roll the tender leaves and scrape the green matter, later they defoliate the entire leaves.

Biological control of Leaf folder and Leaf rollers

- ☆ *Chrysocharis* sp. (Eulophidae) and *Cotesia* sp. (Braconidae) parasitize larvae of *Caloptilia tiselia,* a common tiny leaf folder defoliating cashew.

Diseases of Cashew

Damping off seedlings (*Fusariumspp*, *Phythium*, *Phytophthora palmivora*, *Cylindrocladium scoparium*)

The fungi attack either root or collar region of the tender seedlings.

P. palmivora attack, affected seedlings become pale, showing water soaked girdles of dark tissues around the stem and seedlings droop and die. During the rainy season, this disease can devastate a nursery in few days if proper control measures not used. Initial symptoms are water soaked leaf spots, which progresses to necrotic brown to dark brown color spots. Infection reaches the whole seedling quickly causing its death (Freire *et al.,* 2002).

C. scoparium wilting and withering of seedlings and presence of necrotic lesions leading to rotting of underground parts occur. The disease occurs in the nursery stage, causing mortality of seedlings.

Management

Drenching of seedlings/grafts at fortnightly intervals with Bordeaux mixture 1% or Blitox 3 g/lit of water during rainy season and cloudy weather conditions.

Anthracnose (*Colletotrichum gloeosporiodes*)

This one of the most important disease in cashew plantations caused by *Colletotrichum gloeosporiodes*. During the rainy period the disease reaches its highest severity, spreading quickly inside and between plants. Temperatures ranging from 22°C to 28°C and at least 10 h of saturation are excellent conditions for the infection.Lesions produced by *pathogen* on young leaves of cashew seedlings or adult plants are initially water-soaked and become orange-brown to light-reddish with age and sporulation of the fungus (Freire *et al.,* 2002). It can attack leaves, twigs, inflorescences, young apples and fruits. In severe cases leaves and fruitlets become totally blighted and drop. Under suitable conditions the disease can reduce yield by up to 50%, drastically affecting apple and kernel quality (Cardoso *et al.,* 1994).

An inflorescence and young shoot blight presumed to be caused by combined attack by tea mosquito (*Helopeltis atonii*) and fungi has been shown to be primarily caused by the tea mosquito bug. The fungi were only saprophytic colonizers (Nambiar and Brahma, 1979).

Management

- ☆ Removal of effected leaves.
- ☆ Monitor for the disease particularly during rainy season and cloudy weather conditions which enhance the disease very fast.

Gummosis (*Lasiodiplodia theobraomae*)

- ☆ In older plants the disease may take several months before the firastsymptos can be detected.
- ☆ Gum exudation is the most visible symptoms after yellowing and leaf drop.
- ☆ It produces dark cankers and cracks on the trunk or stem portion. Lesions may reach deep woody tissues completely blocking the sap flow.
- ☆ Exudation of resin-like gum from the trunk or stem region.
- ☆ Pathogen obstructs in water and nutrient transport in the plant causing defoliation, yellowing and die back symptoms on tree canopy.
- ☆ Severe gummosis leads to death of the plant.
- ☆ Grafted cashew seedlings can be affected by the fungus *Lasiodiplodia theobraomae* during the healing of graft joints. The pathogen can be either endophytic in cashew root stock or scion portion or even be disseminated to seedlings through the nursery men knives. (Freire *et al.,* 2002)

Management

- ☆ Clean cultivation
- ☆ Removal and destruction of dead trees

- ☆ Application of biofortified Trichoderma in the pits at the time of replanting
- ☆ Application of biofortifiedTrichoderma during rainy season (2kg of Trichoderma species + 10 kg neem cake + 90 kg FYM) to all the trees in the garden every year.

References

Brazilian cashew kernals. Mycopathologia **145**: 95-103.

Cardoso, J.E., Cavalcante, M.J.B., Cavalcanti, J.J.V. and Vieira, A.C.F., 1994. Identification of Sources of Resistance in dwarf clones of cashew to important leaf diseases. Proceedings of the 13 th annual meeting of the Brazilian society of fruit crops. Salvador- BA, Brazil (Abstract,123) .

Cashew pest database. ICAR-DCR. https://cashew.icar.gov.in/pestsite/

Elakkiya, E., Paramasivam, Sivaraj., Arumugam. and Vijayaprabhakar. 2017. Growth and Performance of Cashew Nut Production in India- An Analysis. *International Journal of Current Microbiology and Applied Sciences*. 6. 1817-1823.

F.C.O Freire, J.E Cardoso, A.A dos Santos, F.M.P Viana. 2002. Diseases of cashew nut plants (Anacardium occidentale L.). Brazil, *Crop Protection*. **21**(6):489-494.

Haritha Paul and Ushadevi K. N. 2022. The Trend in Area, Production, Productivity of Cashew Nut in India with Special Reference to Kerala, *Asian Journal of Agricultural Extension, Economics & Sociology,* **40**(3) : 1-8.

Manual on Integrated pest management in cashew.2016. Director, ICAR-Directorate of Cashew Research.

Nambiar, K.K.N. and Brahma, R.N. 1979. Important diseases of cashew nut plants (*Anacardium occidentale* L.) in Brazil and their control. Crop Protection **21**:489-494.

Pillai, Rajmohan and Santha, P. 2008 "The World Cashew Industry," *Rajan Pillai Foundation*, Kollam (Kerala) India.

Satyagopal, K., S.N. Sushil, P. Jeyakumar, G. Shankar, O.P. Sharma, D.R. Boina, S.K. Sain, Ram Asre, R. Murali, Sanjay Arya, Subhash Kumar, Suresh D. Ekabote, S. Lingaraju, V.K. Koshta, A.K. Awasthi, R.N. Ganguli, M.P. Thakur, A.S. Kotasthane, K. Rajashekarappa, Poonam Srivastava, K. P. Singh, Surajit Khalko, Nripendra Laskar, Ayon Roy and T.K. Hath. 2014. AESA based IPM package for Cashewnut. pp 42.

Stonedahl, G. L. 1991. The oriental species of *Helopeltis* (Heteroptera: Miridae): A review of economic literature and guide to identification. *Bulletin of Entomological Research*, 81: 465-490.

Stonedahl, G. M., Malipatila, M. B. and Houstona, W. 1995. A new mirid (Heteroptera) pest of cashew in northern Australia. *Bulletin of Entomological Research*, 85 (2): 275.

Vanitha, K. 2015. Insect pests of cashew and their management. ICAR- Directorate of Cashew Research, Puttur, Karnataka, India.

Vanitha, k., Raviprasad, T. N. and Veena, G. L. 2022. Management of minor pests of cashew and pollinators of cashew ICAR-DCR, Puttur pp. 95-101.

Vanitha. K. and Raviprasad, T.N. Notes on Tea Mosquito Bug Infesting Cashew Plants and its Management. *Indian Farmer* **7**(04): 287-291; April-2020.

INSECT PESTS AND DISEASES OF CASHEW

Apple and nut borer
(Thylacoptila paurosema)

Apple and nut borer
(Thylacoptila paurosema)

Cashew stem and root borer
(Plocaederus ferrugineus)

Cashew stem and root borer
(Plocaederus ferrugineus)

Cashew stem and root borer
(Plocaederus ferrugineus)

Cashew stem and root borer
(Plocaederus ferrugineus)

Leaf and blossom webber,
Lamida (Macalla) moncusalis

Leaf and blossom webber,
Lamida (Macalla) moncusalis

Leaf and blossom webber,
Lamida (Macalla) moncusalis

Leaf and blossom webber,
Lamida (Macalla) moncusalis

Leaf miner
Acrocercops syngramma

Leaf miner
Acrocercops syngramma

Leaf miner
Acrocercops syngramma

Shoot tip cater pillar
Anarsia epotias

Gummosis:
Lasiodiplodia theobraomae

Gummosis:
Lasiodiplodia theobraomae

Anthracnose
(Colletotrichum gloeosporiodes)

Damping off in Cashew seedlings

Chapter-8

ARECANUT DISEASES, INSECT PESTS AND THEIR BIO MANAGEMENT

N.B.V. Chalapathi Rao, A. Kireeti and B. Neeraja

Dr YSRHU-Horticultural Research Station

Dr. YSR Horticultural University

Ambajipeta, Dr. B. R. Ambedkar Konaseema District, A.P.

Introduction

Arecanut (*Areca catechu* L.) is an important crop among plantation crops, cultivated in many parts of the world with an area of 846 (000 ha) production of 1.21 mt and productivity 1.43 (tons/ha). Mostly cultivated in countries like India, China, Bangladesh, Indonesia, Malaysia, Srilanka, Philippines and New Guinea etc. Though the production of arecanut is confined in some areas, commercial products are majorly shared across the globe and approximately 700 million people regularly chew betel nut worldwide thus support the livelihood of millions of small and marginal farmers and traders (Guo *et al.*, 2020; Peng *et al.*, 2015). In India cultivation of arecanut is from time immemorial whereas, the area increased by 70% over the past two or three decades and the production boom changed. In India, arecanut is cultivated in an area of 518,000 ha with an annual production of 853,000 metric tonnes (NHB 2019). Arecanut is famous for masticatory motive and used to chew with betel leaves or as scented supari and its cultivation is mainly focused in the South-Western and North-Eastern area. The states of Karnataka, Kerala, Assam, West Bengal and Meghalaya are the principal producers and account for greater than 70% of the area and production (NHB, 2019).

It is a perennial crop is vulnerable for attack by a diverse group of diseases, insect pests and disorders throughout the year at every stage of its life. Important diseases and insect pests are detailed below.

Diseases

Bud rot and Crown rot (*Phytophthora meadii*)

Among the many reported diseases, *Phytophthora* diseases results in partial or total crop loss in individual palms or death of the palm itself. The fungus infects the immature tender nuts, growing bud and the crown resulting in fruit, bud and crown rots respectively (Saraswathy, 2004). Identification of *Phytophthora* infecting arecanut palms was started by imperial mycologist Butler (1906) about 115 years back. Later, quite a lot of systematic research has been done on *Phytophthora* diseases and their management. An attempt has been made in this article to review the work done so far on *Phytophthora* diseases of arecanut and to find out the gaps in research to identify future research stratums and priorities for effective disease management.

Occurrence and distribution of diseases: In India, the occurrence of the disease was increased year after year due to the raise in inoculums within the endemic regions of South India inflicting extreme economic loss. In the endemic areas of India, the disease occurs straight away after the onset of the monsoon rains and prevails till mid-August due to congenial weather conditions. Crown rot disease occurs due to extended infection of pathogen and prevails till October due to prolonged rains and congenial weather factors. Systematic surveys conducted by various researchers from CPCRI (Anonymous 2015, 2016, 2017) indicated the occurrence of the disease severity from 5 to 28.6% and varied from garden to garden within major arecanut growing districts of Karnataka and Kerala and distributed based on the amount and pattern of rainfall. Among the disease-endemic areas, the highest incidence of fruit rot was observed in Karkala of Udupi district (41.25%) followed by Sirsi of Uttara Kannada (38.80%) districts of Karnataka (Anonymous 2014). From the above findings, concluded that-*Phytophthora* diseases of arecanut are widely distributed in major arecanut growing regions of India mainly in Karnataka, Kerala, Tamilnadu, Andhra Pradesh parts of Andaman Islands and Meghalaya and subsequently the occurrence of *P. meadii* was reported from different areca growing areas of Kerala and Karnataka (Dutta and Hegde 1987; Santhakumari and Hegde 1987; Saraswathy 1994).

The *Phytophthora* can infect arecanut seedlings or adult palms and the symptoms can be seen on buds, spear leaf, leaf sheath, mature leaves, inflorescence and fruits, but the frequent infection appears on immature tender nuts during the monsoon season which is commonly called as "fruit rot". The infection on the bud leads to "bud rot" and on the crown region through the leaf sheath leads to complete "crown rot".

Fruit rot or Koleroga or Mahali (*Phytophthara meadii*)

Fruit rot disease is characterized by rotting and heavy shedding of immature nuts (Butler 1906). A detailed account of the disease with its causal organism was documented by Coleman (1910) in his pioneering work on fruit rot. The appearance of disease could be identified by the unusual shedding of fruits during south-west monsoon season. The symptom appears as dark green water-soaked lesions at the nut surface usually close to the perianth. The pathogen makes entry to the host tissue through the stomata or epidermis. The entry is aided by the mycelium or germ tube of the germinating sporangia (Coleman, 1910). Under the laboratory conditions, the infection initiates within 18–19 h on the wounded tissues or occurs 4–5 days.

Anabe roga or foot rot (*Ganoderma lucidum*) Symptom initiates as yellowing of outer whorl of leaves gradually extending to inner whorls. In advanced stages, the leaves droop and drop off, finally leaving only one or two leaves along with spindle. Rotting of roots and internal tissues of the basal portion of the stem are other symptoms. It is difficult to identify the diseased palm in the early stages of infection.

Inflorescence dieback (*Colletotrichum gloeosporioides*) the symptom appears as yellowing and drying of rachis from the tip towards the base followed by shedding of female flowers.

Leaf spot (*Colletotrichum gloeoesporioides*; Phyllosticta**)** The disease appears during South West monsoon season. Plants up to 10 years are more susceptible. Small brown to dark brown or black round spots are seen on the lamina. Severe infection causes stunted growth seedlings.

Yellow leaf disease This is caused by Phytoplasma and transmitted by the vector, *Proutista moesta*. Yellowing of leaves is the main symptom. Yellowing starts from the tip of the leaflets of the outer leaves and sometimes seen in the middle whorl. Yellowing spreads gradually extending from the margin to the middle of lamina; portions near the midrib remain green. In advanced stages, yellowing spreads to all leaves completely and they dry and falloff. Kernel of the nuts of affected palms become soft, show blackish discolouration and assumes a spongy texture.

Biological Control of Diseases on Arecanut

To counteract the ill effects of fungicides, efforts were made to identify effective bio-agents and plant extracts against *P. meadii*. The laboratory study indicated that *Trichoderma viride, T. harzianum, Aspergillus terreus, Pseudomonas fluorescence* etc. are potential bioagents against *P. meadii* (Saraswathy 2000b). Similarly study conducted at CPCRI, Kasaragod, indicated the compatibility of bacterial isolates antagonistic against *P. meadii* with Potassium phosphonate and Ovis 20 fungicides (Anonymous 2003). While screening, three native *Trichoderma* species against a highly virulent isolate of *Phytophthora meadii*, *T. virens* exhibited the highest mycelia growth inhibition (62.5%) of *P. meadii* both in simultaneous inoculation and inoculation of *Trichoderma*, 48 h after inoculation of *P. meadii* (Prathibha et al. 2016). Hence it was suggested that these effective fungicides and *T. virens* could be evaluated in field conditions to develop effective integrated management practices for *Phytophthora* diseases. Result of multilocational evaluation of microbial consortia of *Trichoderma harzianum*, *Pseudomonas fluoroscens* and *Bacillus megatarium* revealed a significant reduction in disease severity and development of new roots, increase in number of leaves and yield of palms (Gangadhara Naik *et al.* 2019). The biocontrol agents also effective in reducing the nut fallen, increase in number of root as well as population of bioagents multiplied in the soil. May be because of involving in mycoparasitism, antibiotic and competition for nutrients also induce the defense responses in plants. *Pseudomonas fluorescens* also used for soil borne pathogens it will help in root colonization. *Trichoderma harzianum* (IMI304056) @ 20 ml (105×10-7 cfu/ml stock soln/L water+ *Pseudomonas fluorescens* (NCIB9046) @ 20 ml (105×10-7 cfu/ml stock soln/L water+ *Bacillus megatarium* (NCTC9848) @ 20 ml (105×10-7 cfu/ml stock soln/L water+ soil application of microbial consortia @ 100 g/plant along with 1 kg of FYM/decomposed compost to the soil per year) was found effective in reducing the disease incidence significantly reduced number of nuts fallen in treatment. Among the various plants extracts tested, 10% aqueous extract of leaves of Mehendi (*Lawsonia inermis*) and sacred basil (*Ocimum sanctum*) were effective in checking the growth of *P. meadii* under In vitro condition (Saraswathy 2002). Antifungal activity of 13 plant extracts tasted Invitro, garlic bulbs and rhizome of haldi (Turmeric) were highly inhibitory to *P. arecae* and was followed by leaf extracts of Mehendi (*Lawsonia inermis*), Neem, Tulsi, and Tridax (Ningdale et al. 2018). Under pot culture study, areca seedlings were artificially infected with *P. meadii* and were sprayed with individual plant extracts and results revealed that the extracts of *Andrographis paniculata* followed by *Eucalyptus globules* were effective in checking *P. meadii* infection on areca seedlings (Anonymous, 2004). The above studies indicated that the source plant of extracts could be used for soil suppression of *Phytophthora* population to below the level of infection especially in areca plantations where black pepper and cocoa are grown as mixed crops.

The studies conducted by various researchers on antagonists against Ganoderma isolate of arecanut revealed that, fungal bio control agents like *Trichoderma harzianum* (65.00 to 72.20 % inhibition), *T. viride* (59.86 to 66.0 % inhibition), *T. koningi* (59.86 % inhibition) and *Pencillium* sp. (42.0 % inhibition) over control under dual culture technique. Among the bacterial bioagents, *B. amyloliquefaciens* (Ba - 43) (63.93 % inhibition) followed by *B. subtilis* (Bs - P22) (59.27 per cent inhibition) *P. putida* (Pp - 1) (70.03 % inhibition) over control in dual culture technique (Modh Zainudin and Abdullah 2008, Iyer *et al.*, 2004, Bhansali 2003, Bhaskaran 1990 and Srinivasulu *et al.*, 2001). The efficacy of strains may vary based on Volatile Organic Compounds (VOC) produced by microorganisms played an important role during their evolution in the context of their interactions, community population and functional dynamics.

Biological control of inflorescence die back in arecanut can be effective by removing the fully affected inflorescences and destroy them by burning to prevent the spread and severity of the disease. Bio agents like *Trichoderma virens* and *Trichoderma viride* A study conducted by Nidhina and Hegde 2013, on biological and chemical agents together on the growth of *C. gloeosporioides* causing inflorescence die back in arecanut at the Department of Plant Pathology, CPCRI, Kasaragod revealed that the combination of bioagents with fungicides (*Trichoderma* sp. viz., *Trichoderma virens* and *Trichoderma viride* and fungicides *viz.* Blitox 50 W and Mixol 72) provided higher disease suppression than achieved with fungicides and bioagents when used alone. To reduce the release of chemical pesticides to the environment, integrated control strategies have been adopted extensively by combining both bioagents and chemical agents.

The following measures can be taken to control the disease Yellow leaf disease: Proper management of the garden is the best way to check the occurrence of the disease. Improve the drainage. Apply 2kg neem cake per palm per year. Phytosanitary measures like cutting and burning of the dead palms along with bole and roots should be followed strictly. Isolate the diseased palms by taking trenches of 30cm wide and 60cm deep around the palm.

Insect Pests

Mites (red and white) (*Raoiella indica* and *Oligonychus indicus*). Mites feed on lower surface of arecanut leaves. The colony is found under white webs. Leaf shows yellow speckles and bronzed appearance. The attack is severe in summer months.

Spindle bug (*Carvalhoia arecae*) Symptom appears as linear dark brown necrotic lesions on spindles and the opened leaves show these lesions as patches.

Root grub (*Leucopholis burmeisteri*) Grubs feed on growing roots. Infested palms show a sickly appearance with yellowing of leaves, tapering of stem and reduction in yield.

Distribution: Dominant species of white grubs are *Leucopholis lepidophora* mainly distributed in the hilly areas of India, whereas, *Leucopholis burmeisteri* and *Leucopholis coneophora* distributed throughout the costal areas of Karnataka and Kerala. The third instar grubs feed on tender roots of areca nut and cause damage and expressing symptoms like yellowing of leaves, stem tapering, reduced length of internodes, reduced vigour, nut fall leads to a reduction in yield and ultimately death of the palm. Kalleshwaraswamy *et al* (2015). reported that *L. lepidophora* caused 27.86-36.97 per cent damage to areca nut with a yield loss of 39.79–41.60 per cent in different districts of Western Ghats of Karnataka. Third instar grubs feeding on tips of tender roots of areca nut. High population density it denuded the entire base of the bole and six to eight grubs are enough to kill palms.

Scale insect (*Aanidiella arientalis and Ischnaapsis langinastris*) The scale insects are seen feeding on nuts, rachillae and leaves. The damage is done by sucking the sap from the plant tissues. As a result of continuous sucking, the tissues become yellow in colour and severe feeding leads to withering and shedding of buttons/ fruits. Damage is very heavy during drought conditions.

Lady bird beetles, *Chilacarus nigrita* and *Chilacarus circumdatus* are found to be effective biocontrol agents against the scale insects attacking arecanut. These can be released in affected areca garden to control the scale insects.

Inflorescence caterpillar (*Tirathaha mundella*) The pest causes considerable economic loss to the crop by directly feeding on the tender rachillae and female flowers. Mechanical injury is a pre-disposing factor for infestation. Affected spadices may be force opened and if the female flowers are damaged, the inflorescence should be removed and burnt. If damage is partial, remove affected portion. The infestation is manifested as bore holes on the inflorescence with webbed or damaged rachillae and female flowers.

Biological Control of Insect Pests on Arecanut

The mite can be controlled by predators like Phytoseiidae family, *Amblyseius channabassavanni, Amblyseius longispinus* and *Amblyseius raoiellus* ((Daniel, 1979 & Gupta, 2003), Coccinellidae family: *Stethorus keralicus* (Daniel,1976; Puttaswamy and Rangaswamy 1976; Daniel 1979) *Stethorus tetranychi* (Daniel,1979) *Stethorus parcepunctatus* (Daniel, 1979) *Jauravia soror* (Daniel, 1979) *Spilocaria bisettata* Dermestidae family*: Aspectes indicus (*cited in Daniel, 1979*) Areca catechu* as reported by researchers surveyed in Kerala and Karnataka of India.

Aspergillus candidus is an entomo-pathogen fungus, produces carcino toxins controlls spindle bug of areca palms, Care should be taken while implementing this practice for maintenance of the gardens against spindle bugs Dhileepan *et al.,* 1990.

The imposition of aqua formulation of entomopathogenic nematode (EPN), *Steinernema carpocapsae* (CPCRI-SC1) @ 60 lakh IJs/palm during June - July followed by combined application with insecticide imidacloprid 17.8 SL @ 0.0045% during September - October was caused 96.6% and 92.5% reduction in grub population in farmers field at Koluru and Harakere village respectively over pretreatment grub population followed by EPN alone application of S. carpocapsae caused 74.3% and 79.1% and talc formulation of *H. indica* caused 54.8% and 51.7% reduction in grub population in the respective villages as reported by (Gangadhara Naik *et al.,* 2019). Pheromone trapping could be an essential method in integrated pest management strategy for management of areca nut white grub. White grub, *L. lepidophora* is challenging to manage because it remains in the soil for the whole period of its life cycle except for a short period adult beetles emerged for mating Adarsha *et al.,*2019.

The following measures can be adopted for controlling the root grub pest: Provide good drainage. Collect adult beetles in the evening hours of 6.30-7.30PM, after 8-10 days of premonsoon showers and kill them. Apply organics like neem cake (2kg / palm / year) for improving the soil structure and thereby enabling the regeneration of roots.

Brachymeria nephantidis Gahan (Hymenoptera: Chalcididae) and *Elasmus puctulatus* Verma and Hayat (Hymenoptera: Eulophidae), were reported as parasitoids of *T. mundella* by saneera *et al.* 2019.

Conclusion and Future Prospects

The fungal and bacterial bio-control agents are found to be potential bioagents for the integration in the Integrated Disease Management approaches to combat many pest and diseases in Areca nut like Fruit rot or Koleroga or Mahali, Foot rot disease, Bud rot and Crown rot, mite, spindle bug etc,. There are many research findings on biological control of Areca nut crops if implemented in a proper and systematic way helps in developing the sustainable eco friendly environment, develops awareness among the farmers on importance of the biocontrol agents and are using the biocontrol programme throughout the cropping programme.

References

Adarsha, S. K., Kalleshwaraswamy, C. M. and Shivanna, B. K. 2019. Reproductive status of areca nut white grub, *Leucopholis lepidophora* Blanchard (Coleoptera: Scarabaeidae). *Journal of Entomology and Zoology Studies*, **7**(3): 544-548.

Anonymous. 2014. Annual Report 2013–14. Central Plantation Crops Research Institute, Kasaragod, p 48.

Anonymous. 2016. Annual Report 2015–16. Central Plantation Crops Research Institute, Kasaragod, p 53.

Anonymous. 2017. Annual Report 2016–17. Central Plantation Crops Research Institute, Kasaragod, p 49.

Bansali, R. 2003. Rejuvenation of Ganoderma affected khejri trees through bio control agents BCAs success story. *Div III CAZRI, Jodhpur*. 2003.

Bhaskaran, R. 1990. Biological control of Thanjavur wilt disease of coconut. *In: National Symposium on Bio control of Root Disease*. Annamalai University, Annamalai nagar. Proceedings: pp. 7–8.

Butler, E. J. 1906. Some diseases of palm. *Agric J India* 1:299–310.

Coleman, L. C. 1910. Diseases of the areca palm. 'Koleroga', Mycological Series. *Bulletin No. 2,* Department of Agriculture, Mysore State, pp 92.

Daniel, M. 1976. Biology and predatory habits of the lady bird beetle *Stethorus keralicus* Kapur (Coleoptera: Coccinellidae), predatory on the palm mite. *Journal of Plantation Crops*, **4**: 7-9.

Daniel, M. 1979. Survey of the indigenous predators of arecanut phytophagous mites. In: Venkata, C.S. (ed). Proceedings of the Second Annual Symposium on Plantation Crops. Plant protection (entomology, microbiology, nematology, plant pathology and rodentology). Indian Society for Plantation Crops, Kasaragod, 555 pp.

Dhileepan, K., Nair, R. R.and Leena, S. 1990. *Aspergillus candidus* Link, as an entomopathogen of spindle bug, *Carvalhoia arecae* M & C (Miriidae; Heteroptera). *Planter*, **66**(775): 519-521.

Dutta, P. K. and Hegde, R. K. 1987. Studies on two Phytophthora diseases (Koleroga of arecanut and black pepper wilt) in Shimoga district, Karnataka State. *Plant Pathol News* **15**:12–29.

Gangadhara Naik, B., Maheswarappa, H. P., Rajkumar, M., Kalleshwaraswamy, C. M., Gowdra, Nagamma. and Lata, S. 2019. Evaluation of entomopathogenic nematodes for the management of white grub, *Leucopholis lepidophora* Blanchard (Coleoptera: Scarabaeidae). *Journal of Entomology and Zoology Studies,* **7**(1): 09-13.

Gupta, S. K. 2003. A monograph on plant inhabiting predatory mites of India. Part 2: Order: Mesostigmata. Memoirs of the Zoological Survey of India. Zoological Survey of India, Kolkata, **20**(1):1-185.

Iyer, Rohini., Parvathy, Meera., Lekha, G., Hegde, Vinayak. and Gunasekharan, M. 2004. Management of basal stem rot disease of *Areca catechu* L. in India. *J. Plant. Crops* **32**(1): pp. 25-27.

Kalleshwaraswamy C. M., Adarsha, S. K., Naveena, N. L. and Sharanabasappa. 2015. Incidence of arecanut white grubs (*Leucopholis* spp.) in hilly and coastal regions of Karnataka, India. *Current. Biotica.* **8**(4):423-424.

Mohd Zainudin., N. A. I. and Abdullah, F. 2008. Disease suppression in Ganoderma - infected oil palm seedlings treated with *Trichoderma harzianum. Plant Prot. Sci.,* **44**: 101-107

Puttaswamy and Rangaswamy, H. R. 1976. *Stethorus keralicus* Kapur (Coleoptera: Coccinellidae), a predator of the areca palm mite. *Current Research* **5**: 27– 28.

Saneera, E. K., Chandrika Mohan., Shreevihar, Santhosh. and Shivaji, H. Thube. 2019. First report of two chalcidoids parasitizing arecanut inflorescence caterpillar, *Tirathaba mundella* Walker (Lepidoptera: Pyralidae) from Karnataka, India. *Journal of Plantation Crops,* **47**(2): 124-127.

Santhakumari, P. and Hegde, R.K. 1987. Studies on Phytophthora diseases of plantation crops. *Plant Pathol News.* l 5:1–2.

Saraswathy, N. 1999. Annual report. Central Plantation Crops Research Institute, Kasaragod, pp 96.

Srinivasalu, B., Vijay Krishna Kumar, K., Aruna, K. and Rao, D.V.R. 2004. "Bio control of major pathogens of coconut". *J. Plant. Crops* **32**: 309-313.

Chapter-9

BIOLOGICAL APPROACHES IN FLOWER CROPS

A. Snehalatha Rani[1], T. Nagalakshmi[2] and D. Srinivasa Reddy[3]

[1]Dr YSRHU-Horticultural Research Station
Dr. YSR Horticultural University
Kovvur, East Godavari District, A.P.
[2]Dr YSRHU-Horticultural Research Station
Dr. YSR Horticultural University
Venkataramannagudem, West Godavari District, A.P.
[3]Dr YSRHU-Citrus Research Station
Dr. YSR Horticultural University
Tirupati, Tirupati District, A.P.

Introduction

Insect pests, diseases together with weeds and postharvest losses accounts to 40-48 percent of yield losses in agriculture. Floriculture is not an exception to the attack of these biotic factors and production of quality flowers of international standards has become a major challenge in commercial floriculture because of pest problems. Diseases and pests are of prime importance causing significant losses in floriculture and require continuous monitoring and time to time implementation of control measures. Various control measures are being implemented across the world against the pests and diseases. However, chemical fungicides play a major role in combating the problem. Use of chemicals is not sustainable as they contaminate food, water, air, soil, and leads to development of resistance in pests. Biological control is an eco-friendly and effective alternative means for reducing the pest and diseases through the use of natural enemies. Predatory and parasitic insects along with pathogens forms the three means of biological control in agriculture. Based on their biological origin and their mode of action, the predators and parasitoids are classified as entomophagous or macrobiological agents and pathogens (fungi, bacteria, viruses and nematodes) are classified as entomopathogens or microbiological agents.

Biological control methods are being applied against several pests such as red spider mite, aphids, thrips, mealybug, white fly, etc. in the production of flower crops particularly rose and chrysanthemum. Species of *Gliocladium, Trichoderma, Bacillus, Streptomyces*, and non-pathogenic *Fusarium* are being implemented in the management of soil borne pathogens of floriculture such as powdery mildews, root rots, downy mildew, grey mould and *Fusarium* or *Verticillium* wilts. There are two types of biological control: Conservative and Augmentative. Conservative biological control involves cultivation and maintenance of diversity that will enhance natural enemy population of pests. The practices involve intercropping and/or the introducing of appropriate botanical species that can be maintained in or around the main crop, increase in plant diversity to provide shelter and food (such as nectar, pollen, and honeydew) to the natural enemies etc.

Augmentative biological control involves the production of natural enemies in a large scale at the laboratory and aiming their release in the field. Even though both the methods have been implemented in modern pest management systems, augmentative biological control has been found successful in commercial floriculture particularly in United States, Brazil, Canada etc. These countries realized the importance of biological control in commercial floriculture especially under protected cultivation and has been implementing the same to combat long standing pest and disease problems. Canadian growers are leaders in the use of biological control in floriculture greenhouses. Ninety percent of growers have used some form of bio-control at one time or another. More than half of the floriculture producers in USA currently use or have used biological controls in their floriculture production system (Schumacher, 2002). There are quite a few studies that successfully demonstrated that biological control strategies are feasible under commercial greenhouse conditions for bedding plants biological control strategies are feasible under commercial greenhouse conditions for bedding plants (Opit *et al.*, 2009).

Floriculture in India

Floriculture in India occupies 305 thousand hectares with a production of 2301 thousand tonnes loose flowers and 762 thousand tonnes cut flowers. (National Horticulture Database, Second Advance Estimates; 2019-20) Ornamental flowers are now commercially cultivated in several states with Andhra Pradesh (19.1%), Tamil Nadu (16.6%), Madhya Pradesh (11.9%) surpassing other producing states like Karnataka, West Bengal, Mizoram, Gujarat, Odisha, Jharkhand, Haryana, Assam and Chhattisgarh.

Floriculture in India includes both open farm conditions as well as protected conditions. It comprises mainly rose, tuberose, gladiolus, chrysanthemum, orchids, anthurium, carnations, marigold etc. India's total export of floriculture was Rs. 575.98 Crores/77.84 USD Millions in 2020-21 (Kumarasamy and Harshavardhini, 2021). The area under flower crops got almost tripled from 106000 ha during 2001-02 to 339000 ha during 2018-19. Similar trend was also noticed in production of flowers in India. The major importing countries were U S A, Netherland, U Arab Emts, U K, and Germany.

Insect Pests and Diseases of Ornamental Crops

Insect pests

Rose

California red scale (*Aonidiella aurantia*) belonging to Hemiptera: Diaspididae, Thrips (*Scirtothrips dorsalis* Hood and *Rhipiphorothrips cruentatus* Hood belonging to Thysanoptera: Thripidae, Aphid (*Macrosiphum rosae* L.) belonging to Hemiptera: Aphididae, Bud borer (*Helicoverpa armigera* Hubner) belonging to Lepidoptera: Noctuidae, Whitefly (*Bemisia tabaci*) belonging to Hemiptera: Aleyrodidae, Tobacco caterpillar (*Spodoptera litura*) belonging to Lepidoptera: Noctuidae, Spider mite (*Tetranychus cinnabarinus* Boisd. and *Tetranychus urticae* Koch) belonging to Acari: Tetranychidae are the major insect pests of rose reported from India.

Gerbera

Whitefly (*Trialeurodes vaporariorum* (Westwood), *Bemisia tabaci* Guen. (Hemiptera: Aleyrodidae), Leaf miner (*Liriomyza trifolii* Burgess) (Diptera: Agromyzidae), Thrips (*Thrips palmi* Karny) (Thysanoptera: Thripidae), Aphid (*Myzus persicae* Sulzer) (Hemiptera: Aphididae), Tobacco caterpillar (*Spodoptera litura*) (Lepidoptera: Noctuidae), Two-spotted spider mite (*Tetranychus ludeni* Zacher) (Acari: Tetranychidae), Cyclamen mite (*Phytonemus pallidus*) and slugs and snails are the major insect pests of gerbera crop in India.

Carnation

Thrips, *Frankliniella schultzei* Trybom. (Thysanoptera: Thripidae), Bud borer, *Helicoverpa armigera* (Hubner) (Lepidoptera: Noctuidae), Two-spotted spider mite, *Tetranychus urticae* (Acari: Tetranychidae) are the three insect pests associated with carnation crop in India.

Orchids

There are five predominant species of scale insects *viz.*, ti scale, *Pinnaspis buxi*; Florida red scale, *Chrysomphalus aonidum*; Lecanium scale, *Lecanium* sp; Soft brown scale, *Coccus hesperidum* and Boisduval scale, *Diaspis boisduvalli* which damage orchids round the year. Two aphids, yellow aphid, *Macrosiphum luteum* and black aphid, *Toxoptera aurantia*, shoot borer, *Peridaedala* sp. (Lepidoptera: Tortricidae), Thrips, *Dichromothrips nakahari* (Thysanoptera: Thripidae), Mealybug,

Pseudococcus sp., Red spider mite, *Tetranychus urticae* Koch (Acari: Tetranychidae) and Snail, *Cryptaustenia verrucosa* and *C. heteroconcha* are the major problems in Orchids.

Chrysanthemum

Aphid, *Macrosiphoniella sanborni* Gillette., *Myzus persicae* (Sulz) (Hemiptera:Aphididae), Thrips, *Microcephalothrips abdominalis* (Crawford), and *Frankliniella* spp. (Thysanoptera: Thripidae), Leaf miner, *Liriomyza trifolii* (Burgess) (Diptera: Agromyzidae) and mites are the insect pests of significance in chrysanthemum.

Anthurium

Whitefly, *Aleurotulus anthuricola* (Hemiptera:Aleyrodidae) and Thrips, *Chaetanaphothrips orchidii* (Moulton) (Thysanoptera:Thripidae) are the pests of significance.

Tuberose

Leaf caterpillars, *Polytela gloriosa* F. and *Brythyscrini* Fb. (Lepidoptera:Noctuidae), Aphids, *Rhopalosiphum nymphaease* Linn. and *Myzus persicae* (Sulzer). (Hemiptera: Aphididae), Lily bulb mites, *Rhizoglyphus echinopus* (Fum and Rob) and *R. robini* (Acarina: Acarididae) are the various pests of tuberose in India.

Biological Control of Insect Pests on Flowers

Aphids

Aphids are the most serious pests of floricultural crops. Except anthurium and carnation, the pest is a major problem in all other flower crops in India.The bodies of aphids are oval and soft, and they are 1.5-3.0 mm tall. Even though they don't cause major losses, they decrease aesthetic value of the crop, often cluster in large colonies of feeding on young flower buds and tender unfolding leaves. Severe infestation may weaken pedicles and distort young shoot and leaf growth (Hoffer *et al.*, 2000).

Biological control of aphids

Predators and entomopathogens are the two major groups of biological agents that are successful against aphids throughout the world. However, entomopathogens are the most effective agents when compared to predators. Entomopathogens also have an added advantage of host specific nature and are harmless to other beneficial insects and non-target organisms. *Lecanicillium lecanii* is the most widely used entomopathogen against aphids across the world (Alavo *et al.*, 2001) and in India. Aphid population may be brought down by spraying 1% neem or pongamia oil or spraying *Lecanicillium lecanii* at 3.0 g/l during evening hours (Prem Jit Singh *et al.*, 2015). *Beauveria bassiana* another entomopathogen used for aphid control (Sayed *et al.,* 2019).However, the disadvantage with the use of entomopathogens is their requirement of high humidity for establishment in host tissue.

Coccinellids, Chrysopids and Cecidiomyiids are the successful predator groups reported to feed on aphids. They are coccinellids, insects belonging to miridae family,*Macrolophus caliginosus* , *Aphidolates aphidimyza, Adalia bipunctata* and *Aphidius colemani* belonging to cecidomyids etc. Coccinellids constitute the predator group that has been most studied as the natural enemies of aphids.

Aphidiidae and Aphelinidae families of the Hymenoptera group constitute the parasitoids of aphids (Wei *et al.,* 2005).

Leaf miner

Leaf miner, *Liriomyza trifolii* (Burgess) (Diptera: Agromyzidae) is a serious problem on cut chrysanthemum and gerbera. The mature leaf miners are 1.3-2.3 mm in length and coloured grey-black. The female insect makes small cuts on the leaves, feed on juice sap and lay eggs. After hatching, larva mine into the leaf and form characteristic serpentine tunnel in the leaf. The damaged leaves turn yellow, dry and then fall off. Development of young plants and seedlings is affected because of larval and adult leaf miner feeding (Hore *et al.,* 2017).

Biological control of leaf miner

The successful biological agents against leaf miners across the world are parasitoids. They are *Diglyphus isaea, Chrysonotomyia chlorogaster, C. formosa, Hemiptarsenus zilahisebessi* and *H. varicornis.*

Diglyphus isaea is one of the most important parasitoid effective against leaf miner reported from Mediterranean region. It is a small, black, non-stinging wasp that searches out leaf miners and lays its eggs. The wasp kills the leaf miner as it lays its egg, and the emerging larvae use the dead miner as food. The adult parasite exits the tunnel sometime later and will then go on to lay 200 – 300 eggs during its nearly month-long life. *D. isaea* is an efficient parasitoid of *L. trifolii, L. bryoniae and L. huidobrensis* larvae from Liriomyza species. As the leaf miner escapes pesticide treatments because they are protected inside the tunnels, the biological control agent, *D. isaea* is one of the best alternatives in control of leaf miner problem in chrysanthemum and gerbera. However, *D. isaea* is an ectoparasite on leaf miner larvae and is more difficult to handle and transport than the braconids, which are endoparasites and emerge from leaf miner pupae(Woets and Van der Linden, 1985).

Two-spotted spider mite

Mite is a major insect problem on flower crops. Spider mites,*Tetranychus cinnabarinus* Boisd and *Tetranychus urticae* Koch in rose, two-spotted spider mite *Tetranychus ludeni* Zacher and Cyclamen mite *Phytonemus pallidus* in gerbera, two-spotted spider mite, *Tetranychus urticae* in carnation, orchids, jasmine and chrysanthemum are few of them. Sucking of plant sap by the insects makes plants yellow and leads to reduced plant growth. Red spider mites usually stay infest leaf and flower parts, and they multiply very fast in hot weather (Blindeman and Van Labeke, 2003).

Biological control of mites

Many natural enemies are reported against red spider mite across the world. However, predator mites are the commercially available natural enemies of spider mite. Predator, *Phytoseilus persimilis* is reported against the pest from mediterranean region and especially feeds on egg and nymph stages of the pest. Life cycle of *Phytoseilus persimilis* is shorter time than spider under the same temperature that makes the predator a potential bio-control agent. Further, the reproductive rate of the predator is also higher, a desirable character in bio-control strategy (Huffaker *et al.,* 1969).

Flower Thrips

Thrips are one of the ubiquitous pests on flower crops. The pest is reported in all the flower crops. *Scirtothrips dorsalis* Hood and *Rhipiphorothrips cruentatus* Hood in rose, *Thrips palmi* Karnyin gerbera, *Frankliniella schultzei* Trybom in carnation, *Dichromothrips nakahari* in orchids, *Microcephalothrips abdominalis* (Crawford), and *Frankliniella* spp. in chrysanthemum, and *Chaetanaphothrips orchidii* (Moulton) in anthurium are the reported thrips species in India.Thrips are very small insects and the body size ranges 1-1.3 mm in length. They cause severe damage on leaves, petals and flowers. They reduce the aesthetic nature of the flowers and severe infestation leads to floral abnormalities.

Biological control of flower thrips

Orius aldipennis, O. laevigatus, O. majusculus, and *O. niger* are some of the biological control agents of *F. occidentalis*(Bosco and Tavella, 2013). Predator mites *Neoseiulus (Amblyseius) cucumeris* is effective against *F. occidentalis* in greenhouses. *Thrips tabaci* do not cause important threat compared to the damage caused by *F. occidentalis.* In general, the natural enemies controlling *F. occidentalis* control *T. tabaci* as well.

Mealybugs

Mealybug, *Pseudococcus* sp. is an important pest on orchids in India. Two mealybugs, *Maconellicoccus hirsutus* and *Planococcus citri* infest *Hibiscus rosasinensis* L. Adult female is slightly oval, 3-5 mm in length, and are cream, straw yellow, light pink in colour. The white colour of mealy bug is due to the white waxy layer covering the body. The spread and infection of mealy bugs happens when larvae hatch out of the eggs. Pest infestation leads to poor development of flowers and may drop off. In addition to direct damage, they also act as vectors for some of the viruses.

Biological control of mealybug

Important bio-control agents against mealybug are coccinellid predator, *Cryptolaemus motrouzieri* Muls., green lacewing *Chrysoperla carnea* (Stephens) and the hymenopteran parasitoid, *Leptomastix doctiilopii* and *Leptomastidea abnormis*. *C. montrouzieri* has been in use worldwide against mealy bugs (Gkounti *et al.*, 2015; Messelink *et al.*, 2016). Both larval and adults feed on mealy bug population effectively. The predator is voracious and feed on 1000 eggs or 300-500 nymphs of mealy bug (Mani *et al.*, 2011). Life cycle of the predator is completed in 30 days. It has been multiplied in large scale under controlled conditions and successfully controlled mealy bug under field conditions. *C. montrouzieri* effectively controlled both *P. citri* and *M. hirsutus* population on hibiscus under field conditions. *C. carnea* is a general predator with a wide climatic adaptability (Senior *et al.*, 2006). Both the species of *Leptomastix* are effective biological control agents against *P. citri* on ornamental plants in protected conditions of Netherlands.

Whitefly

Whitefly is an emerging pest of significance in ornamental crops grown under protected cultivation. Whitefly lays eggs on the underside of the leaf that hatch in 10 days. They remain and feed on the leaf for another 10 to 15 days. It has a broad host range infecting 249 plant genera. Whitefly is considered a major pest under protected conditions. Whitefly while sucking the sap from phloem, excretes honeydew at all developmental stages. Secreted honeydew attracts the saprophyte sooty mould fungi and severe sooty mould on leaf surface reduces photosynthesis and thereby hindering respiration. Aesthetic value of the crop is also highly affected.

Biological control of whitefly

Parasite, *Encarsia formosa* is effective bio-control agent against whitefly in horticultural crops. *Encarsia Formosa* has been a successful bio-control agent and is being used across the world after its commercialization(Speyer, 1927).The wasp can discriminate between parasitized and healthy hosts, thus avoiding superparasitism. The parasite can be easily mass reared at the lab conditions with minimum cost and facilities. The parasites are introduced into glasshouses as pupae; this is a resilient stage protected by the larval skin of the whitefly. Introduction of *E. formosa* when greenhouse whitefly densities are low is of paramount importance for successful control. An initial density of 10 adult *T. vaporariorum* per 100 m^2 is considered too high (Ekbom, 1977). A primary parasitization rate of 50% or more is necessary to prevent an increase in the population of *T. vaporariorum*.

Diseases

Rose

Black leaf spot (*Diplocarpon rosae*) (Imperfect state: *Marssonina rosae*), Powdery Mildew (*Podosphaera pannosa*) (=*Sphaerotheca pannosa*), Stem canker/ Dieback (*Diplodia rosarum, Botryodiplodia theobrome, Colletotrichum gloeosporoides, Coniothyrium fucklelii*), Blossom blight (*Botrytis cinerea*), Downy mildew (*Peronospora sparsa*), Crown gall (*Agrobacterium tumefaciens*), Mosaic (Rose mosaic virus, Rose ring pattern virus) are the major diseases affecting rose flower production (Prem Jit Singh *et al.*, 2015).

Gerbera

Leaf spot caused by *Alternaria gerbericola, A. alternata, A.tenis, A.dianthi*, Bacterial leaf spot caused by *Pseudomonas cichorii*, Botrytis blight by *Botrytis cinerea* are the foliage pathogens of Gerbera whereas Phytophthora crown rot caused by *Phytophthora cryptogea*, Pythium root rot caused by*Pythium irregulare*, Rhizoctonia crown rot by *Rhizoctonia solani*, Fusarium foot/crown rot by *Fusarium oxysporum* are the soil borne diseases leading to crop death. Powdery mildew caused by *Golovinomyces cichoracearum (=Erysiphe cichoracearum)*, *Podosphaera fusca (=Sphaerotheca fusca)* is a common disease on gerbera along with viral diseases, Tobacco rattle, gerbera mosaic, impatiens necrotic spot and cucumber mosaic virus. (Prem Jit Singh *et al.*, 2015).

Carnation

Fusarium stem rot/wilt caused by *Fusarium oxysporum f. sp. dianthi, Alternaria blight by Alternaria dianthi* and *Alternaria dianthicola*, *Rhizoctonia* stem/collar rot caused by *Rhizoctonia solani*, Rust by *Uromyces dianthi*, *Phythophthora* blight by *Phytophthora parasitica, P. capsici, P. cryptogea*, Bacterial wilt caused by *Pseudomonas caryophyllii* and viruses are the major diseases of carnation in India (Prem Jit Singh *et al.*, 2015).

Orchids

Black rot/ Crown rot/ Heart rot by *Pythium* sp. and *Phytophthora* sp., Anthracnose caused by *Gleosporium/ Colletotrichum* sp., Blossom blight/ Petal blight caused by *Botrytis* sp. Bacterial soft rot caused by *Erwinia* sp. and viruses are the diseases observed on orchids (Prem Jit Singh *et al.*, 2015).

Chrysanthemum

Leaf spot (*Septoria chrysanthemella* and *S. obesa*), Stemphylium/Alternaria ray speck (*Stemphylium* and *Alternaria* spp.), Verticillium wilt (*Verticillium alboatrum*), Powdery mildew (*Erysiphe cichoracearum*), Blossom blight (*Botrytis cinerea*) and Stem rot (*Rhizoctonia solani*) are the major diseases observed on chrysanthemum in India (Prem Jit Singh *et al.*, 2015).

Anthurium

Anthracnose caused by *Colletorichum gloeosporioides*, Root Rot caused by *Calonectria crotalariae, Pythium splendens, P. spinosum, P. vexans*, and *Phytophthora* spp., *Rhizoctonia* spp. and *Fusarium* spp., Bacterial blight, *Xanthomonas dieffenbachiae, X. campestris. pv. dieffenbachiae, X. axonopodis pv. dieffenbachiae*, are the few diseases on Anthurium (Prem Jit Singh *et al.*, 2015).

Tuberose

Bulb and scale rot caused by *Fusarium oxysporum f. sp. lilii*, Rhizoctonia rot caused by *Rhizoctonia solani*, Foot rot, *Phytophthora cactorum, P. nicotianae, P. parasitica*, etc., Root rot, *Pythium ultimum*, grey mold, *Botrytis elliptica*, viruses like Lily symptomless, CMV, Tulip breaking, Tobacco rattle, Tobacco ring spot are the major diseases in the crop (Prem Jit Singh *et al.*, 2015).

Biological Control of Diseases on Flowers

Powdery mildew

Powdery mildew is a serious problem in rose (*Podosphaera pannosa) (=Sphaerotheca pannosa)*, gerbera *(Golovinomyces cichoracearum =Erysiphe cichoracearum, Podosphaera fusca = Sphaerotheca fusca)* and chrysanthemum*(Erysiphe cichoracearum).*

Podosphaera pannosa is the most devastating disease on roses worldwide. Both garden roses as well as greenhouse roses, gerbera and cut chrysanthemums. Disease initially appears as a whitish powder covering entire foliage, stems and bud. Predominant symptoms appear on upper side of the leaves but also affects the under side of leaves and other plant parts including stems, young shoots, flowers buds, and young fruits. Severe infection reduces the photosynthetic area of the foliage, increase respiration and transpiration finally leading to loss of vigour and yield. Severe losses between 20 to 40 per cent were reported under favourable climatic conditions.

Powdery mildew fungus is an obligate parasite and can entirely depend on the plant for its survival. It's one of the destructive and widely distributed plant pathogens across the horticultural crops. Pathogen infects leaves, twigs and floral parts of the plants. Severely infected leaves bend and harden. Leaves often become slightly red and covered with a white mycelial growth. Buds may also shrink and do not bloom (Pasini et al., 1997).

Golovinomyces cichoracearum (*Erysiphe cichoracearum*) causes powdery mildew in gerbera and chrysanthemum crops. Moderate temperatures with alternating cool nights and warm days are congenial for the development of the pathogen. Disease may aggravate under closed canopy and poor air circulation. Powdery mildew pathogen is highly specialized and can adopt the conditions favourable to the host plant. Most of the mycelium remains outside the plant parts but sinks root-like structures called haustoria into plant cells to obtain nutrients. Severe and prolonged incidence greatly reduce the plant vigour and production. Grayish white, powdery spots develop on leaves. These may be scattered at first but can quickly cover entire leaves giving them a frosty appearance. All aboveground plant parts can become infected including the flower stalk and petals.

Biological control of powdery mildew

Several microorganisms (Table 1), have been found to be effective as bio-control agents against powdery mildew fungi. Mycoparasitism by *Ampelomyces* spp. is the best known example of biological control of powdery mildews. However, the affectivity of mycoparasites on powdery mildew pathogen depends on the isolate of bio-control agent and environmental conditions. Powdery mildews pose a greater difficulty to the biological agents in a way that they thrive well under dry conditions where as most of the biological control agents require relative humidity above at least 70%.

Table.1: Powdery mildews in ornamentals and their biological control agents

S. No	Crop	Causal organism	Bio-control agent	References
1.	Rose	*Podosphaera pannosa* (=*Sphaerotheca pannosa*)	*Ampelomyces quisqualis*	Kiss, 2003 and Verhaar et al., 1999b
2.			*Aphanocladium album*	Verhaar et al., 1999b
3.			*Pseudomyza flocculosa*	Belanger et al., 1994; Hajlaoui and Belanger, 1991
4.			*Pseudomyza rugulosa*	Hajlaoui and Bélanger, 1991 and Verhaar et al., 1999b
5.			*Tilletiopsis minor*	Verhaar et al., 1999b
6.			*Tilletiopsis pallescens*	Ng et al., 1997
7.			*Tilletiopsis washingtonensis*	Hajlaoui and Belanger, 1991
8.			*Verticillium lecanii (Zimm.) Viegas*	Verhaar et al., 1999 and Kiss, 2003
9.			*Gliocladium catenulatum*	Janice et al., 2011
10.			*Trichoderma harzianum*	Janice et al., 2011
11.			*Bacillus subtilis*	Janice et al., 2011
12.	Gerbera	*Golovinomyces cichoracearum* (=*Erysiphe cichoracearum*),	*Bacillus subtilis*	Moyer and Peres, 2008
		Podosphaera fusca = *Sphaerotheca fusca*	Phosphorus acid	Moyer and Peres, 2008
		Podosphaera fusca = *Sphaerotheca fusca*	Potassium bicarbonate	Moyer and Peres, 2008
13.	Chrysanth-emum	*Erysiphe cichoracearum*	Ammonium oleate	Michael Bradshaw, 2015

*Ampelomyces quisqualis*is commercialized as water disperasable granules under the trade name AQ10. The other well commercialized bio-control agents against powdery mildew of rose are *Gliocladium catenulatum*, *Bacillus subtilis* and *Trichoderma harzianum* (Sztejnberg *et al.*, 1989; Janice *et al.*, 2011).

The yeast-like fungal antagonists effective against powdery mildew fungus are *Pseudomyza* and *Tilletiopsis* species. *Pseudomyza rugulosa,Pseudomyza flocculosa* and *Tilletiopsis pallescens* are the best bio-control agents against powdery mildew of rose. The *P. flocculosa* is a species which has a higher colonization ability, increased production of antibiotics and greater adaptability to the dry environmental conditions suitable for the host. Bio-control agent does not penetrate the host mycelium but induces a rapid plasmolysis of the host cells (Hajlaoui and Belanger, 1991; Hajlaoui *et al.*, 1992). The mechanism of bio-control activity of *P. flocculosa* is confirmed as production of antibiotics through culture filtrate assays and isolation of fatty acid molecules. The synthesized fatty acid molecules from the culture filtrate imitated the bio-control mechanism(Avis *et al.*, 2000). The high sterol content in the cell membrane facilitates *P. flocculosa* to avoid its own antibiotics. There will be no chance of pathogenic resistance as the antibiotics degrade very rapidly in nature.It is observed that *P. flocculosa* controlled powdery mildew pathogen on par to chemical control with dodemorphacetate and microfine sulfur (Belanger *et al.*, 1994). Further the treatment improved flower quality when compared to fungicidal sprays. *Tilletiopsis* spp. are another phyllosphere yeasts effective against powdery mildew by the production of antibiotics (Elad et al., 1996).*T. pallescens* reduced 78-94 per cent conidia formation of rose powdery mildew (Ng *et al.*, 1997).

In a study involving all the bio-control agents against powdery mildew, *Verticillium lecanii* showed better control of rose powdery mildew when compared to *Ampelomyces quisqualis, Aphanocladium album, Pseudomyza rugulosa* and *Tilletiopsis minor*. Bio-control ability of *V. lecanii* involves the mechanism of hyperparasitism as well as antibiosis (Verhaar *et al.*, 1999b; Kiss, 2003).

The different bioproducts and botanicals found effective against powdery mildews include anhydrous milk fat (AMF) and soybean oil (SBO) emulsions (Chee *et al.*, 2011), foliar application of compost tea (Seddigh *et al.*, 2014) and the extract of grape fruit at 0.066 per cent concentration (Wojdyla, 2001).

The other tested biological control agents includes *Bacillus megaterium, Bacillus pumilus; Bacillus subtilis; Brevibacillus brevis; Bacillus thuringiensis; Pantoea agglomerans; Pseudomonas marginalis; Pseudomonas fluorescens.*

Grey mould

Grey mould is an important disease on flower crops under protected conditions. It infects rose, gerbera, orchid species, chrysanthemum and tuberose crops. Grey mould causes substantial losses in greenhouse flowers, during storage and transit. The disease initiates as water-soaked lesions on petals which later turn to brown. The surface of the flowers is covered with intense grey mycelia and spore mass under high humid conditions. The pathogen enters through injuries or mechanical damage on the tissues. Latent infections on the flower can also cause losses during storage and marketing process.

Biological control of grey mould

Environment plays equal part in biological control of Botrytis blight. Removal of senescent flowers, dead and decaying tissue reduces the colonization of Botrytis blight. As high relative humidity favours the pathogen, venting and heating at sunset to reduce the formation of dew on roses overnight reduces the disease pressure. In addition, horizontal air movement around plants to reduce relative humidity has been effective on some ornamental crops. In some cases, application of supplemental calcium has been found effective against Botrytis blight in both cut flower production and post-harvest storage.

Ulocladium atrum has been found to be antagonistic to *B. cinerea* suitable for application on necrotic above ground plant tissues. The bio-control agent has high competitive colonization ability with superior ecological competence. Long persistence of viable conidia on leaf surface, high resistance to interruptions of leaf wetness and the strong antagonistic activity over a broad temperature range are the strong features of *Ulocadium atrum* for its success against the pathogen over broad range of hosts such as pot roses, Pelargonium, strawberries and onion (Kohl, 2004).

Other biological agents against grey mould include *Trichoderma inhamatum* and *Gliocladium roseum* which effectively reduced sporulation of the pathogen in rose leaf residues (Elad *et al.*, 1993). Combined treatment with yeasts and *Bacillus subtilis, Bacillus subtilis, C. cladosporioides*, and *C. oxysporum* are some other effective agents (Elad *et al.*, 1993; Tatagiba *et al.*, 1998).

Ethylene inhibitors are another alternative to reduce pathogen infection by retaining the membrane integrity of plant tissues and delaying senescence. Ethylene inhibitors: aminoethoxyvinylglycine (AVG), aminooxyacetic acid (AOA), 1-methylcyclopropene (1-MCP)and silver thiosulphate (STS) significantly reduced grey mould disease severity (Elad 1988a; Seglie et al. 2012).Combination of AVG and 1-MCP was found most effective in suppressing *B. cinerea* infection in cut 'Pink Beauty' roses under simulated transport conditions.

Application of nano silver to prolong vase life of cut flowers by suppressing ethylene synthesis is also tested and found most effective against *B. cinerea* infection (Ha et al. 2021; Park et al. 2017).Nano silver releases silver ion, that can replace the cofactor Cu2+ for the ethylene receptor and thereby effectively suppresses ethylene binding and downstream reactions resulting in the inhibition of the ethylene responses in rose petals (Ha *et al.*, 2021). Silver ions also releases ROS causing damage to nucleic acids, lipids, and proteins of the cells of *B. cinerea*. AVG, 1-MCP, and NS are promising alternatives for grey mould disease control of flower crops particularly under protected conditions (Ha *et al.*, 2021).

Downy mildew

It has a widespread impact in the greenhouses which are not ventilated in autumn and spring when night temperature is low. The disease is seen especially on the leaves and branches of young plants, on the pedicle, bud and sepals. Shapeless stains, which are coloured differently from magenta to black, are observed on the leaves. The leaves become yellow and drastic defoliation is seen in the end. The underside surfaces of the leaves are full of spores in humid environments. As opposed to this, the upper surface of the leaf is full of with spores in powdery mildew (Hoffer *et al.*, 2000).

Biological control of downy mildew

Bacillus subtilis was found effective against rose downy mildew disease.

Fusarium species

Fusarium stem rot/wilt caused by *Fusarium oxysporum f.* sp. *dianthi* in carnation, root or crown rot caused by *Fusarium oxysporum* in gerbera, root rot by *Fusarium* spp in anthurium and bulb and scale rot caused by *Fusarium oxysporum f.* sp. *lilii* in tuberose are of significant diseases in India. The pathogen affects and damage ornamental plants at all production stages. Disease symptoms include yellowing, stunting, and death in seedlings and older plants. Wilting and drying of the entire plant is the characteristic symptom of Fusarium infection in plants. Besides, early symptoms include yellowing and drying of the lower leaves followed by wilting. Xylem tissues turn brown leading to plant death. In carnation and gladiolus, the initial symptoms may be one-sided.

Fusarium oxysporum is a well-known soil borne pathogen that infects plant roots or wounded cuttings. It survives in the soil as thick-walled, dormant chlamydospores for longer periods. When suitable host is available, chlamydospores germinate in response to exudates from plant roots. Hyphae then penetrate the roots, colonize the cortex, and xylem tissue. Infection of the vascular system obstructs water and nutrient absorption by the plant leading to wilting symptom. Pathogen produces sporodochia on which millions of micro and macro conidia are produced. The spores are airborne and infect nearby soils and plants. After plant death, fungus produces chlamydospores that are released into the environment as the tissue decomposes.Pathogen spread by soil, wind, water, infected cuttings, and contaminated tools, equipment and clothing.

Biological control of *Fusarium* species

Management of Fusarium wilt is very difficult. It's a soil borne disease and a combination of physical and biological methods can be adopted for successful control. Physical methods include fumigation and solarisation of the soil. Avoidance and exclusion are must to restrain the pathogen infecting healthy soils. Antagonistic microorganisms effective against *Fusarium* spp include *Trichoderma harzianum*, *Streptomyces griseoviridis*, *Streptomyces lydicus*, *Pseudomonas fluorescens* in cut flowers.

Fusarium wilt can be effectively managed by biological control means. For example, suppressive soils for Fusarium wilt of carnation were identified throughout the world. The suppressive soils against Fusarium wilt of carnation in the United States reported *Pseudomonas* and *Alcaligenes* species whereas non-pathogenic *Fusarium* species were reported from France and Italy. Bio-control agents reported do have high rhizospheric competence along with antagonistic activity (Garibaldi *et al.*, 1989). Strains of *Trichoderma viride* showed good antagonistic activity against Fusarium wilts and have been used in chrysanthemum (Singh and Kumar, 2011). Commercial preparation of beneficial microbes was only marginal in suppressing Fusarium wilt on gladiolus (Elmer, 2006b), but when combined with fungicides, they provide superior control of Fusarium wilt of cyclamen than fungicides alone (Elmer and McGovern, 2004). Microbial optimization of the nutrient solution through the addition of beneficial microorganisms has been successfully attempted for *Phytophthora cryptogea* on gerbera (Garibaldi *et al.*, 2003).

Alternaria species

Alternaria leaf spot is a disease of significance on cut flowers like rose, gerbera (*Alternaria dianthi*), and dianthus. Blight or rot at leaf bases and around nodes, which later girdle the entire infected stem or flower stalk leading to die back symptom. It produces ashy white spots that are covered with dark brown to black fungal growth. Infection on leaves causes constriction and twisted leaf tips and the tip may be killed.

Biological control of *Alternaria* species

Bacillus subtilis and *Streptomyces griseoviridis* are the two biological agents found effective against the leaf spot.

Black spot

It is one of the important diseases on rose caused by *Diplocarpon rosae*.Initial symptoms appear on upper side of the leaf as brown to black spots with fringed margin. With disease progression, the spots are surrounded by a yellow halo that later turn complete yellow. In severe conditions, several spots coalesce and forms bigger spots. These black spots along with yellowing leads to premature leaf fall. Disease first appears on mature or older leaves of the cane and progresses upwards. Defoliation starts at the base and proceeds to younger leaves.

Biological control of black spot

Trichoderma viride and *Pseudomonas fluorescens* are found effective against *D. rosae*. The bio-control agents in addition to antagonism, induced increased synthesis and activity of defense-related enzymes and chemicals in plants such as phenylalanine ammonia lyase (PAL), peroxidase (PO), polyphenoloxidase (PPO) and total phenolics. (Karthikeyan *et al.*, 2007).

Crown gall

Crown gall in rose is caused by *Agrobacterium tumefaciens*, a soil inhabiting bacterium (Smith & Townsend 1907) Conn 1942.The bacterial pathogen produces large galls at the base of the stem or on roots. With disease progression, galls may be formed higher on stems. Galls are soft compared to surrounding plant tissues. Young plant if infected early may not bloom. Generally, galls are formed at or just below the soil surface in the basal or root crown region. They may frequently appear on roots but less frequently on stems and leaves.

Biological control of crown gall

Biological control has been very successful against crown gall of rose and other stone crops (New and Kerr 1972; Kerr 1980) by using *Agrobacterium radiobacter* K84. *A. radiobacter* is a nonpathogenic agrobacterium species belonging to biovar 2 subgroup. Planting material dipped in *A. radiobacter*K84 suspension at 10^7 to 10^8 cells per ml provided better control of crown gall disease. *A. radiobacter*K84 secretes agrocin 84 which is an antibiotic-like substance that specifically inhibits many tumorigenic *Agrobacterium* strains.

Other biological agents against crown gall disease include *Bacillus megaterium, Paenibacillus polymyxa, Pseudomonas fragi, Pseudomonas viridiflava, Pseudomonas asplenii, Curtobacterium flaccumfaciens* and *Curtobacterum* sp (Tolba and Soliman, 2012)

Conclusion and Future prospects

In the current scenario, biological control agents provide a better alternative to control pests and diseases in ornamental crops. There are many success stories on biological control of ornamental crops if implemented in a proper and systematic way. Augmentative biological control has been found successful in commercial floriculture particularly United States, Brazil, Canada. Farmers in these countries are all aware of the importance of the bio-control agents and are using the bio-control programme throughout the cropping programme. Further studies on characterization of bio-control agents, elucidating their mode of action by utilizing molecular techniques, viability and survival of the bio-control agents, improving their shelf life and field evaluation, adaptability of the bio-control agents to adverse environmental conditions are needed. Latest technologies like nanotechnology may be used in preparation of biopesticides, which stabilizes bioproducts with controlled release and precise doses. Utilizing artificial intelligence tools to monitor the crop and pest status, identifying the areas with pest infestation, helps in optimizing the decision-makings. Application of appropriate biological agent for each target-pest at the ideal time, at the most susceptibility place, and in exact amounts further improves the biological control programme.

References

Alavo, T. B. C., Sermann, H. and Bochow, H. 2001. Biocontrol of Aphids Using Verticillium Lecanii in Greenhouse: Factor Reducing the Effectiveness of the Entomopathogenic Fungus', *Archives of Phytopathology and Plant Protection*, **34**(6): 407 – 424.

Avis, T. J., Boulanger, R. and B′elanger, R. R. 2000. Synthesis and biological characterization of (Z)-9-heptadecenoic and (Z)-6- methyl-9-heptadecenoic acids, fatty acids with antibiotic activity produced by *Pseudozyma flocculosa. J. Chem. Ecol.* **26**:987–1000.

Belanger, R. R., Labbe, C. and Jarvis, W. R. 1994. Commercial scale control of rose powdery mildew with a fungal antagonist. Plant Disease, **78**:420-424.

Blindeman, L., and Van Labeke, M.C. 2003. Control of the two-spotted spider mite (Tetranychus urticae Koch) in glasshouse roses. *Commun. Agric. Appl. Biol. Sci.* **68** (4 Pt A), 249–254.

Bosco, L. and Tavella, L. 2013. Distribution and abundance of species of the genus Orius in horticultural ecosystems of northwestern Italy. *Bulletin of Insectology* 66 (2): 297-307.

Chee, A.A., Wurms, K.V. and George, M. 2011. Control of powdery mildew (*Sphaerotheca pannosa var. rosae*) on rose (*Rosa* sp.) using anhydrous milk fat and soybean oil emulsions. *New Zealand Pl. Protec.* **64**: 195-200.

Ekbom, B. 1977. Development of a biological control program forgreenhouse whiteflies (*Trialeurodes vaporariorum* Westwood) using its parasite *Encarsia formosa* (Gahan) in Sweden. Zeitschrift ftir angewandte *Entomologie,* **84**: 145-154.

Elad Y, Kirshner B, Gotlib Y (1993) Attempts to control Botrytis cinerea on roses by pre- and postharvest treatments with biological and chemical agents. Crop Prot 12:69-73 DOI:10.1016/0261-2194(93)90023-C

Elad Y, Malathrakis NE, Dik AJ. 1996. Biological control of Botrytis-incited diseases and powdery mildews in greenhouse crops. *Crop Prot.* 15:229–40

Elad, Y. 1988a) Involvement of ethylene in the disease caused by Botrytis cinerea on rose and carnation flowers and the possibility of control. Ann Appl Biol 113:589-598 DOI: 10.1111/j.1744-7348.1988.tb03336.x

Gkounti V.T., Savopoulou-Soultani M., Milonas P.G. Intra-and interspecific interactions between *Nephus includes* and *Cryptolaemus montrouzieri. J. Appl. Entomol.* 2015;139:647–653. doi: 10.1111/jen.12208.

Ha STT, Kim YT, Jeon YH, Choi HW, In BC (2021) Regulation of Botrytis cinerea infection and gene expression in cut roses by using nano silver and salicylic acid. Plants 10:1241 DOI:10.3390/plants10061241

Hajlaoui MR, Benhamou N, B′elanger RR. 1992. Cytochemical study of the antagonistic activity of *Sporothrix flocculosa* on rose powdery mildew, *Sphaerotheca pannosa var. rosae. Phytopathology* 82:583–89

Hajlaoui, M.R. and Belanger, R.R. (1991). Comparative effectsof temperature and humidity on activity of three potential antagonists of rose powdery mildew. *Netherland J. Plant Pathol.*, 97 : 203-220.

Hoffer, M., Pscheidt, J., and DeAngelis, J. (2000). Controlling Diseases and Aphids on Your Roses. EC 1520 (USA: Oregon State University), p.4–6.

Hore, G., Chakraborty, A., and Banerjee, D. (2017). *Liriomyza Trifolii* (Insecta: Diptera: Agromyzidae). The invasive alien agricultural pest species of India. *ENVIS Newsletter*. 23 (1), 12–16

Huffaker CB, de Vrie MV, McMurtry JA. The ecology of tetranychid mites and their natural enemies. *Annual Review of Entomology* 1969;14:125-144.

Janice, F.E., Haselhan, C. and Punja, Z.K. (2011). Evaluation of biological control agents for control of botrytis blight of geranium and powdery mildew of rose. *Canadian J. Plant Pathol.,* 33 : 499-505

Kairo, M. *Cryptolaemus montrouzieri* (Mulsant) (Coccinellidae: Scymninae): A review of biology, ecology, and use in biological control with particular reference to potential impact on non-target organisms. CAB Rev. Perspect. *Agric. Vet. Sci. Nutr. Nat. Resour*. 2013, 8

Kiss, L. (2003). A review of fungal antagonists of powdery mildews and their potential as biocontrol agents. *Pest Management Sci.,* 59 : 475-483

Kohl, J. 2004. Biological control of Botrytis spp. by *Ulocladium atrum* through competitive colonization of necrotic plant tissues. Plant Research International. B. V. Wageningen. April 2004. pp 232.

Mani, M., Krishnamoorthy, A. and Shivaraju, C. 2011. Biological suppression of major mealybug species on horticultural crops in India. *J. Hortl. Sci.* Vol. 6(2):85-100

Messelink GJ, Maanen R, Steenpaal SEF, Janssen A (2008) Biological control of thrips and whiteflies by a shared predator: two pests are better than one. *Biol Control* 44:372–379

Messelink, G.J.; Vijverberg, R.; Leman, A.; Janssen, A. Biological control of mealybugs with lacewing larvae is affected by the presence and type of supplemental prey. *BioControl* 2016, 61, 555–565

Michael Bradshaw. 2015. Thesis on Potential Organic Fungicides for the Control of Powdery Mildew on Chrysanthemum x morifolium. University of Washington.

Moyer, C. and N.A. Peres. 2008. Evaluation of biofungicides for control of powdery mildew of gerbera daisy. *Proc. Fla. State Hort. Soc.,* 121: 389 - 394.

N. Kumarasamy and G. Harshavardhini. 2021. Production and Export Performance of Indian Floricultural Sector. *Just agriculture*. Vol.2 Issue-2, OCT 2021

Ng, K.K., MacDonald, L. and Punja, Z.K. (1997). Biological control of rose powdery mildew with the antagonist yeast *Tilletiopsis pallescens*. *Hort. Sci.,* 2 (2) : 262-266.

Opit, G.P., J. Perret, K. Holt, J.R. Nechols, D.C. Margolies, and K.A. Williams. 2009. Comparing chemical and biological control strategies for twospotted spider mites (Acari: Tetranychidae) in commercial greenhouse production of bedding plants. *J. Econ. Ent.* 102: 336-346.

Park DY, Naing AH, Ai TN, Han JS, Kang IK, Kim CK (2017) Synergistic effect of nano-sliver with sucrose on extending vase life of the carnation cv. Edun. Front Plant Sci DOI:10.3389/fpls.2017.01601

Pasini C, Daquila F, Curir P, Gullino ML. Effectiveness of antifungal compounds against 1037 rose powdery mildew (*Sphaerotheca pannosa var. rosae*) in glasshouses. *Crop Prot* 1038 1997;16:251–6

Prem Jit Singh, D. S. Kakade, N. Majumder, V. Sridhar, Girish, K. S., Prabha K., K. P. Singh and Prasanna Holajjer. (2015). Disease and Pest Management in Flower Crops under Polyhouse. DFR Extension Bulletin No. 12. ICAR – Directorate of Floricultural Research, College of Agriculture Campus, Shivajinagar, Pune - 411 005 (Maharashtra), India

Rosenheim, J.A.; Wilhoit, L.R.; Armer, C.A. influence of intraguild predation among generalist insect predators on the suppression of an herbivore population. *Oecologia* 1993, 96, 439–449.

Sayed, S. M., Esmat Farouk Ali and Saqer Sultan Al-Otaibi. 2019. Efficacy of indigenous entomopathogenic fungus, *Beauveria bassiana* (Balsamo) Vuillemin, isolates against the rose aphid, *Macrosiphum rosae* L. (Hemiptera: Aphididae) in rose production. *Egyptian Journal of Biological Pest Control.* 29:19https://doi.org/10.1186/s41938-019-0123.

Schumacher SK. Economic analysis of production and pest management decisions confronting greenhouse growers (PhD dissertation). Kansas State University, Manhattan, KS; 2002

Seddigh, S., Kiani, L., Tafaghodinia, B. and Hashemi, B. (2014). Using aerated compost tea in comparison with a chemical pesticide for controlling rose powdery mildew. *Arch. Phytopathol. & Plant Prote.,* 47 (6) : 658-664

Seglie L, Spadaro D, Trotta F, Devecchi M, Gullino ML, Scariot V (2012) Use of 1-methylcylopropene in cyclodextrin-based nanosponges to control grey mould caused by Botrytis cinerea on *Dianthus caryophyllus* cut flowers. *Postharvest Biol Technol* 64:55-57 DOI:10.1016/j.postharvbio.2011. 09.014

Senior, L.J.; McEwen, P.K. The use of lacewings in biological control. In Lacewings in the Crop Environment; McEwen, P.K., New, T.R., Whittington, A.E., Eds.; Cambridge University Press: Cambridge, UK, 2001; pp. 296–302.

Speyer ER. 1927. An important parasite of the greenhouse whitefly (*Trialeurodes vaporariorum* Westwood). *Bulletin of Entomological Research* 17: 301–308

Sztejnberg, S., Galper, S., Mazar, S. and Lisker, N. (1989). *Ampelomyces quisqualis* for biological and integrated control of powdery mildew in Israel. *J. Phytopathol.,* 124 : 285-295.

Tatagiba, J. D. S., Maffia, L.A., Barreto, R. W., Alfenas, A.C. and Sutton, J. C. 1998. Biological control of *Botrytis cinerea* in residues and flowers of rose (*Rosa hybrida*). *Phytoparasitica.* **26**:8-19 DOI:10.1007/BF02981261.

Verhaar, M.A., Kerssies, A. and Hijwegen, T. 1999b. Effect of relative humidity on mycoparasitism of rose powdery mildew with and without treatments with mycoparasites. *J. Plant Diseases Prote.* **106**: 158-165.

Wei. J. N., Bai, B.B., Yin, T.S., Wang, Y., Yang, Y., Zhao, L.H., Kuang, R.P. and Xiang, R.J. 2005. Development and use of parasitoids (Hymenoptera: Aphidiidae & Aphelinidae) for biological control of aphids in China. *Biocontrol Science and Technology*, September 2005; **15**(6): 533-551.

Woets, J. and van der Linden, A. 1985. First experiments on *Chrysocharis parksi* Crawford (Hym.: Eulophidae) as a parasite for leafminer control (*Liriomyza* spp.) (Dipt.: Agromyzidae) in European greenhouse tomatoes *Mededelingen van de Faculteit Landbouwwetenschappen, Rijksuniversiteit Gent.* **50**: 763-768.

Wojdyla, A.T. 2001. Grapefruit extract activity in the control of rose powdery mildew and black spot. *Meded Rijksuniv Gent Fak Landbouwkd Toegep Biol. Wet.* **66** (2): 167-177.

Chapter-10

NEMATODES IN HORTICULTURAL CROPS AND THEIR MANAGEMENT WITH SPECIAL REFERENCE TO BIOLOGICAL CONTROL

Ch. S. Kishore Kumar and A. Snehalatha Rani

Dr YSRHU-Horticultural Research Station
Dr. YSR Horticultural University
Kovvur, East Godavari District, A.P.

Introduction

Nematodes are the most numerous multicellular, colourless un segmented worms dwelling in all types of habitats on earth successfully parasitizing plant, animals and rarely on humans. The Word Nematode is derived from Greek word 'nema' means thread, and 'oides' means form. They are ubiquitous in nature hiding in soil as free living (including predators) and as a parasitic feeding on plant and its parts. On plants, these small microscopic worms feeds on all parts *i.e.,* roots, stems, leaves, flowers and seeds with the help of protrusible needle like structure in mouth *i.e.,* stylet, a key identification feature as a plant parasitic nematode (PPN).

During recent years, these Plant-parasitic nematodes (PPN's) are posing a major threat to cultivation of both agricultural and horticultural crops across the world. The full extent of nematode damage in plants is under estimated since the growers are often unaware about their presence because of the symptoms expressed are often non-specific and mimic to micro nutrient deficiencies (Siddique and Grundler, 2018). The Damage due to PPN's has been estimated approximately @ 12.3% ($157 billion) worldwide (Singh et al., 2015) which was more significant compared to invasive insects (around US$70 billion; Bradshaw et al., 2016).

In the present scenario across the world, there is an urgent need of more food production including horticultural crops to cater the needs of growing population (FAO, 2017) for which growers were compelled to cultivate crops under protected cultivation in a bigger way. Taking into consideration of the high economical impact caused by plant parasitic nematodes on food production, a large number of novel strategies have been developed for nematode management using chemicals/nematicides. But due to new regulations against chemicals which are harmful both to human health and for the environment (Zhang et al., 2014, 2017), strategies associated with bio control are considered and were proposed to be a safer alternative for the management of PPN's. Therefore, global demand for the control of plant-parasitic nematodes is also increasing for meeting the desired goals undoubtedly.

In this chapter a brief view of nematodes habitat, life cycle and nature of damage and symptoms of damage is presented.

Habitat

Nematodes habitat on all diversified forms of earth *i.e.,* from ocean depths to top of mountains, most of them living freely and few of them feeding on other micro organisms. Free living nematodes forms an important constituent in organic matter recycling of soil. Of all the nematodes known, about 50 % of them lives in marine environment, 25 % in soil or fresh water feeding on bacteria, fungi and other decomposers and 15 % of them are parasites of animals, and only 10 % of them are parasites of plants called as PPN's.

These PPN's are mostly found at 7-20 cm depth in soil mostly around the plant root zone with required optimum temperature of @ 15-30°c. A thin film of water is very much essential and vital element to complete its biological activities and for its survival. Soil moisture of 35-60% of the field capacity, well aerated and drained light soils along with soil pH of up to 7 are the best congenial factors for most of the PPN's for their easy locomotion and better oxygenation.

Morphology

The Plant parasitic nematodes (PPN's) are small, microscopic and were in range from 0.3 to 2.0mm in length with typically vermiform in shape, thread-like and tubular with tapering towards the both ends. *Paralongidorus epimikis* is the only PPN observed with attaining upto1.2cm, the longest PPN observed so far. Males and females generally look alike. The males remain vermiform for the entire life time whereas, the females based on their parasitic nature deviates from normal cylindrical form and becomes swollen and appears like saccate, lemon shaped. The reproductive, excretory and nervous systems were well developed and were located within the pseudocoelom with lacking of special organs for respiration and circulation systems. Nematodes are dioecious *i.e.,* males and females are separate. Males can be distinguished by the possession of spicules and usually bursa near to its posterior end. The oxygen required for the plant parasitic nematode is obtained through its body cuticle. The cuticle plays a vital role for oxygenation.

Life cycle

The Nematode completes its life cycle in six stages - the egg, four larval or juvenile stages (J1 to J4) and the adult (male or female). In plant parasitic tylenchid nematodes, the first stage juvenile moults within the egg itself, and the second stage juvenile (J2) hatches out. It moults further from J3 to J4. The juveniles are almost very much similar to adults except varying in size and development of reproductive system.

The nematodes complete their life cycle starting from egg to forming an adult and depending upon the duration for completion of their life cycle, they were classified as univoltine (completion of life cycle in a year, example- *Heterodera* sp.) and multivoltine- completing many generations in a year, example- *Meloidogyne* sp . Univoltines have usually host specificity and active for the entire host crop season, whereas multivoltine have wide host range. The duration of nematode life cycle *i.e,* egg to adult is highly influenced by temperature, host and other environmental factors.

The ovipositional behaviour of plant parasitic nematodes vary based on their parasitic nature , as most of them lives in soil by laying eggs singly in soil itself (ecto parasitic) , some entering into plant tissue and laying eggs singly within the tissue (endo parasitic) or some times on root surface clumped together in a gelatinous substances secreted by rectal glands (root-knot nematode), excretory system (citrus nematode), while others retaining within the body of the females (cyst nematodes). The fecundity (eggs laid per female) varies from 10-25 (ecto and migratory) to 50-450 (endo /semi endoparasitic).

Spread

Nematode movement in soil is very much limited and hardly they move a few feet on its own in a year. Most of the PPN's spreads through plant propagative material, farm implements/equipments, and irrigation water including plant debris.

Nature of Damage

Plant parasitic nematodes (PPN's) poses a serious threat to a variety of horticultural crops worldwide undoubtedly (Ali *et al.*, 2017), causing damage in different ways ultimately resulting in death of the plant. The damage caused by nematodes can be observed in two ways *viz.*, direct and indirect damage.

Under direct damage, the nematodes feed upon the plant cells clearing the cell cytoplasm giving appearance of necrotic patches resulting in death of the feeding cells. The nematodes after getting entry into plant tissue migrates at intercellular and intracellular level causing injury which can be observed in lesion and burrowing type of nematodes. In some nematodes, they form special feeding sites in plant cells and due to enzymatic action it thus forming giant cells (root knots) and syncytia (cyst nematode) thus disrupting the flow of essential nutrients and water from root to shoot resulting in poor and stunted, finally leading to death of the plant. It is also observed in few nematodes which feed on germinating seedlings resulting in death before pre emergence of the seedlings, *e.g.,* Sting nematode.

In indirect damage, wounds or injury caused by nematodes favors the entry or involvement of other microorganisms in causing disease complex in plants, and such disease complexes brings irreversible bio and physiological changes of the host plant causing overall reduction in the root system making them vulnerable to various biotic and abiotic stresses. Few nematodes act as vectors also in transmitting fungal and bacterial spores which aggravates the disease spread in a short span of time.

Symptomatology

Nematode incidence on host plant can be observed with onset of symptoms both externally and internally. The external symptoms always mimic with micronutrient deficiencies or like physiological disorders.

External symptoms (above ground)-These symptoms are generally non-specific and necessitate careful examination and confirmation of nematode incidence. The above ground symptoms are stunted in growth, yellowing, wilting, die-back, galls formation (cereal crops, flower crops, leaf spots, necrosis and discolouration etc.).

Internal symptoms (below-ground)-The symptoms expressed below the ground level vary based on the nematode genera and confirms the nematode incidence undoubtedly. The symptoms are root galls (*Meloidogyne* spp), root lesions (*Pratylenchus* spp, *Radophilus* spp and *Helicotylenchus* spp etc.) root swellings and root rots (potato rot- *Ditylenchus destructor*, yam –*Scutellonema bradys*).

Parasitism

Plant parasitic nematodes (PPN's) attacks or parasitize host plants which can be categorised into two types, below ground plant parts and above ground parts (stems, leaves, buds & inflorescence etc.)

Below-ground parts: PPN's attacks below ground plant parts such as on roots or other propagating seed material and complete its entire life cycle within the rhizospheric soil periphery to host plant root system. Based on the parasitism, they were again classified into 3 types,

Ectoparasites

In this type, PPN's spend their life cycle with in the soil moving freely and feeding intermittently on root epidermis and on root hairs but never enters into the plant tissue. The eggs were laid singly in soil. They almost feed on particular site for a brief period and then move to next site, hence also called as migratory ectoparasites e.g., Sting nematode, Needle nematode (*Longidorus* sp). Few nematodes under ecto parasitic nature were observed with sluggish in nature also, feeding continuously on a particular site for a longer time with the help of long stylet e.g., sessile nematode (*Cacopaurus pestis*).

Semi endoparasitic- In this type, the nematodes insert its anterior portion of its body into the plant tissues and feeds for quite significant time and come out without establishing any permanent feeding site on host plant. e.g., stunt nematode- *Tylenchorhynchus* sp. In this type also sedentary semi endoparasitic nature was observed with establishing permanent feeding site on the host plant by inserting its anterior body and lays eggs in masses around its body or on root surface embedded with gelatinous substances secreted by females, e.g., Citrus nematode- *Tylenchulus semipenetranns* and Reniform nematode- *Rotylenchulus reniformis.*

Endoparasitic- This type of parasitic nature is very destructive form where in the PPN's almost completely embedded within the host roots with continuously feeding, reproducing and completing many life cycles within the feeding site.

This endoparasitic nature is divided into two types *viz* ., migratory and sedentary. Under sedentary parasitic nature – the nematode keeps on moving within the root tissue causing extensive damage due to cavity formation in cortex and necrosis of the tissue, laying eggs within cortex and were scattered within the tissue. The juveniles and adults only will move out into soil to reinfect the other roots or other underground healthy parts of same plant or the other nearby plants e.g., Lesion nematode (*Pratylenchus* sp.), Burrowing nematode (*Radopholus similis*) etc.

The other is sedentary endoparasitic type, which is considered to be most dangerous and causing huge economical losses to both agricultural and horticultural crops. In this the PPN's after entering or penetrating into root tissue establishes a permanent feeding site within the root tissue and becomes sessile by loosing its power of locomotion. The body of female becomes sac like or lemon shaped, oviposition may be on the root surface or within the body of female. e.g., Root knot (*Meloidogyne* sp.) and Cyst nematode (*Heterodera* and *Globodera* sp.).

Nematode Management under Protected Conditions

Nematode management is far more difficulty than any other kind of pest because of their diversified habitat and mostly attacking the underground parts of the plants. Nematode management aims at maintaining the population at below the ETL, to reduce the economic losses. The extent of crop damage by PPN's is directly correlated with their population levels in the soil.

Great improvement has been brought in for nematode management especially with the use of chemicals, however generally discouraged due to several reasons out of which due to being very expensive along with other several side effects like residue problems, resurgence of nematode population, environment pollution etc. Moreover usage of chemicals /nematicides was recommended to be used only when the nematode population is very high and quick results are demanded. In order to overcome these hurdles, it is essential to develop new control strategies against nematodes especially with bio agents apart from already existing cultural, physical and chemical methods.

Before going in detail about biological strategy against nematode management, awareness on Cultural, Physical and chemical methods also need to be studied.

Cultural methods- crop rotation with non-host crops (*e.g.*, rotation with mustard crop controlled molya disease of wheat caused by *Heterodera avenae*), fallowing (keeping field fallow reduced the root-knot incidence), flooding (killing nematodes by asphyxiation due to low oxygen), raising trap/enemy crops (*crotalaria*-nematodes fails to reach maturity; onion, garlic & *Tagetes*- root exudates having nematoxic compounds), Organic manuring (stimulation of beneficiary soil micro organisms) and lastly with using of healthy planting material, as this is the main source of spread of PPN's.

The adoption of physical method i.e. hot water treatment was proved successfully especially white tip disease of rice by seedlings dipping at 55 ^{0}c has reduced the nematode incidence in south India. The planting material like bulbs, tubers, suckers to specific temperatures can also be standardised for effective management of nematode incidence during their crop growth stage.

Chemical methods- Use of chemicals/Nematicides is very effective method, but was observed with several side effects for which they were generally discouraged. The chemicals/nematicides can be categorised into two, fumigants and non-fumigants. Fumigants are volatile in nature and turn into gaseous phase when applied to soil, and also there is a chance of getting escape of vapours into atmosphere which can be detrimental to flora and fauna.

Latest chemicals under this group are metham sodium, dazomet, which releases methyl isothiocynate which paralyse and cause death of the nematodes. Non-fumigants chemicals are old group of chemicals which acts as AChE (acetyl choline esterase) inhibitors by binding phosphorylation and deactivating them for further reproduction capacity. Some of the effective chemicals under this group are Fensulfothion, Carbofuran and Phorate etc.

Biological Control

The organism that suppresses the pathogen(enemy micro organism) is referred to as the biological control agent (BCA) which can interact directly with the pathogen through antibiosis and competition for nutrients and space or indirectly with the pathogen through systemic acquired resistance (SAR) and induced systemic resistance (ISR) (Xiang *et al.*, 2018).

Biological control of nematodes aims at regulation of nematode population and/or a reduction in nematode damage through the action of organisms antagonistic to them, which occur naturally or through the manipulation of the environment or the introduction of antagonists. However, sensing very potent and effective bioagent is a challenging task for successful management of nematodes. There are two groups of micro organisms that parasitize nematodes *viz* predators and parasites. Several kinds of predators, parasites and pathogens of PPN's exist in the soils and continuous interactions at biological and physiological takes place in nature.

Among predaceous nature of bioagents -trapping fungi, saprophytic nematodes, protozoans *etc*, have been recognised, and, opportunistic fungi, bacteria, viruses etc were recognised as opportunistic parasites of nematodes. For successful biological control, three components should be considered -naturally (bio agents already present at sufficient levels for nematode suppression in that area), induced (stimulation of available bioagent with the help of inciters) and introduced (application of bioagent externally into that area).

Some of the successful bioagents utilised and proven successful in nematode management under protected and open field conditions are,

Bio Management

Fungi

Nematode destroying fungi can be grouped into three types- predaceous fungi, endozoic fungi and opportunistic fungi. Most of the predaceous fungi belong to order- Moniliales which attacks the nematodes with the specialised structures such as sticky traps (*Monacrosporium cionopagum*) or adhesive knobs (*Dacttylaria candida*) parasitizing the PPN's. Due to their survival competency in nature they were not much commercialised. The Fungi with endozoic nature kills the nematode with either adhesive spores or by simple spores sticking to the nematode bodies which are ingested by nematodes. These fungi considered to be seriously potential bio agents. But for the unknown reasons these fungi also were not much commercialised. e.g., *Meria coniospora*. Regarding opportunistic nature of fungi, they are capable of colonising nematode reproductive structures causing diseases on females and laying deformed or unproductive eggs. e.g., *Fusarium, Paecilomyces, Phoma , Verticillium* and *Trichoderma spp* etc. and *Glomus mossae* and *G.fasciculatum* (Arbuscular mycorrhia fungi) were found promising and successful bioagents against PPN'S. In recent studies it was observed that *Paecilomyces lilacinus* caused egg deformation in *Meloidogyne incognita* and cysts of *Globodera pallida* , a serious menace in potato.

These fungi minimises the space and resource-competition, provide higher nutrients and water uptake to the host plant, modifying root morphology by interactions at rhizospheric soils which constitutes an advantage for the plant-growth. Besides this, they also induces resistance against nematodes by activating hormone-mediated (Salicylic and jasmonic acid pathways) plant-defence mechanisms. Additionally, transport of chemical defence components or plant secondary metabolites and different enzymes contributing to enhance plant defense mechanism. With this the fungi as BCAs is a promising and durable biocontrol strategy in agriculture against PPN's.

Bacteria

Another bioagent which has come into prominence recently due to its antibiosis phenomenon is rhizospheric bacteria particularly PGPR (Plant promoting rhizobacteria). PGPR colonize the root surfaces and rhizospheric soil promoting plant growth and controlling phytopathogens. It was also observed with production of inhibitory alleochemicals and induction of systemic resistance (ISR) in host plants. Some of the successful PGPR strains are *Pseudomonas fluorescens, P.putida, Azotobacter.*

Protected Cultivation

Protected cultivation is a process of growing/raising horticultural crops under controlled environmental conditions such as temperature, humidity and light as per the crop requirements. Crops grown under protected cultivation are flowers (Chrysanthemum, Carnation, Gerbera, Rose, Lilium, Orchid, Gladiolus, etc.,) Vegetables(Tomato, Coloured Capsicum-Yellow and Red Bell Peppers, Cucumber, Broccoli, Red Cabbage, Leafy vegetables, Radish, etc.,), fruits (Strawberry Seedling, Fruit Grafting (like Lemon, Citrus, Mango, Pomegranate, Guava, Litchi, etc) and nurseries of vegetables, Flowers, Tissue Culture, and Clonals for forestry. These crops grown throughout India are seriously infested with nematodes such as *Meloidogyne incognita, M. javanica* (root-knot nematodes) and *Rotylenchulus reniformis* (reniform nematode). Nematode problems on all these crops under protected conditions have assumed alarming proportions leading to huge crop losses (up to 80%) in select crops. The nematode infestations exacerbate severity of fungal diseases leading to complete crop losses.

Under this controlled micro climatic conditions there is a high chance of incidence and fast multiplication of pathogenic micro organisms through the year. Among the pathogens, nematode incidence is one the major constraint in protected cultivation where in the population builds up very rapid and reaches to a maximum threshold level within a year and making very difficult to continue further. Symptoms such as chlorosis and stunting appear after sufficient damage is inflicted.

Keeping in view of this tiny hidden potential enemy to protected cultivation, there is an immediate need to formulate eco friendly integrated management strategies and are the technologies recommended and proven successful for nematode management.

1. **Nursery bed preparation-** Making soil into fine tilth by deep ploughing followed by application of FYM @15-20ton/ha enriched with microbial consortium (*Trichoderma viride*@1kg+*Pseudomonas* spp @ 1kg + *Paecilomyces lilacinus* @ 1kg+Neem cake @ 10kg enriched on FYM@1 ton).
2. Soil application of carbofuran @ 0.3g a.i/m^3 at nursery stage and then again second time application @ 1kg a.i/ha, for effective management of root knot incidence (*Meloidogyne* sp).
3. Soil solarisation of nursery beds with a thin transparent polythene sheet (50-100um) for 15-20 days during summers with light wetting of the soil, which was found very effective against *Meloidogyne* spp in vegetables (cucurbits & vegetables)

4. Seed dressing with bioagents *Trichoderma viride / P.fluorescens* @ 10g/1kg vegetable seed, which was found effective against of root knot incidence.
5. Application of organic amendments such as castor cake and neem cake@ 1t/ha against root knot nematodes in vegetable crops.
6. Paring and hot water treatments of banana suckers @ 55⁰C for 20 minutes combined with application of neem cake @ 1kg/plant and carbofuran @ 20g/plant against nematode wilt complex disease.
7. Selection of certified seed/planting material from registered nursery which is free of nematode incidence.
8. Raising of antagonistic plants-Antagonistic plants serves as a tool for biological control for nematodes as they were known to have nematode suppressant properties and are considered to be promising agents in most of the cases. Some of the antagonistic group of plants are Compositae, Cruciferaceae, Liliacea, Fabaceae and Meliaceae.

Compositae *(Tagetes* spp*)*- Due to nematicidal action of root diffusates of *Tagetes* spp, this plant was utilised and successfully managed PPN'S when grown in a appropriate timings.

Tagetes spp	Time of sowing	Against PPN	Target crop
T. erecta	Preceeding crop	Lesion nematode (Pratylenchus penetrans)	Tuber crops
T. patula	Growing for 3-4 months	Root knot nematode (Meloidogyne incognita)	Any vegetable crops.

Cruciferaceae-White mustard and black mustard under *Cruciferaceae* is another promosing group successfully considered against PPN'S due to presences of isothiocyanates in their root diffusates. Inclusion of these crops as cropping sequence also has reduce root knots and other PPN's in both protected and field conditions.

Liliaceae (Asparagus) –Due to presence of nematicidal glycoside in their roots exudates, this has completely inhibited the larval emergence of cyst nematode (*Globodera* spp and *Heterodera* spp).

Fabaceae *(Crotalaria juncea*)- Due to presence of pyrrolizidine esters, suppresses the *Meloidogyne* spp, and also the PPN's cannot further continues its life cycle due to getting embedded within the root tissues there by the load can be reduced in a significant level.

Meliaceae (*Azadirachta indica*)-The neem was found to have multi dimensional qualities for which being used widely in medicines and in Agricultural purpose. The antagonistic nature of neem was due to root exudates containing chemicals-Azadirachtin, Kaempferol, nimbin, nimbidin, nimbidic acid, quercetin and thionemone. These chemicals inhibit larval hatching of *Meloidogyne* spp and *Rotylenchus* spp in most of the vegetables crops.

Stress on Plant parasitic nematodes due to antagonistic crops may further utilised in all types of cultivation both in Agricultural and Horticultural crops.

Nematode infestation symptoms under field conditions

Yellowing of guava plant due to nematode infestaion

Root knots (*Meloidogyne* spp) on guava roots

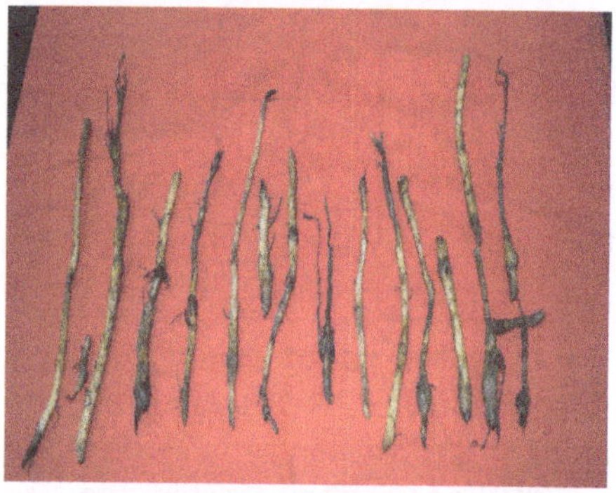

Root galls (*Meloidogyne* spp) on banana roots

Root lesions (*Pratylenchus* spp) on banana roots

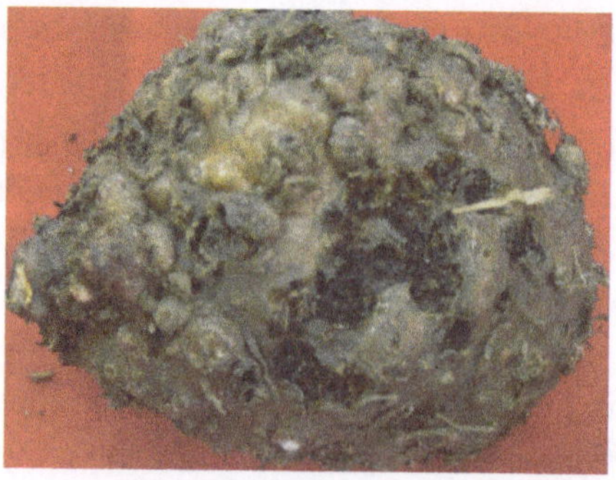

Root knots on Elephant foot yam

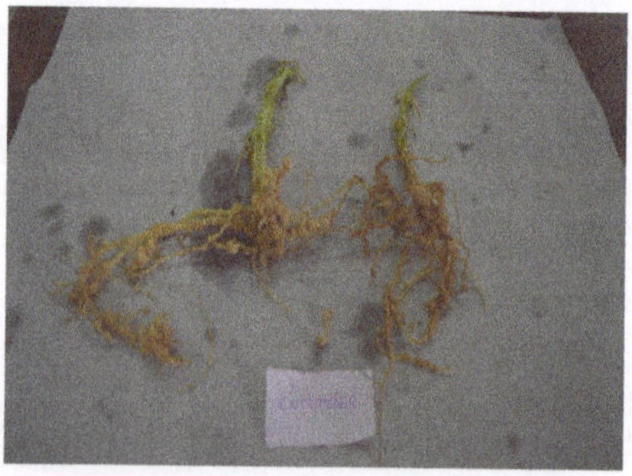

Root galls on cucumber roots

Root knots on tomato

wilting & drying of pomegranate plant due to nematode incidence

Root galls on crossandra

Root galls on sweet potato

Root galls on Ixora (Ornamental plant)

References

Ali, M. A., Anjam, M. S., Nawaz, M. A., Lam, H.-M. and Chung, G. 2018. Signal Transduction in Plant–Nematode Interactions. *Int. J. Mol. Sci.* **19**:1648.

Bradshaw, C. J., Leroy, B., Bellard, C., Roiz, D., Albert, C. and Fournier, A. 2016. Massive yet grossly underestimated global costs of invasive insects. *Nat.Commun.* 7: 1–8.

Navid, Sabir. and R.K.Walia. 2017. Management of Nematodes in Protected cultivation with short notes on key pests- published by Project coordinating cell, IARI, New Delhi.

Shamim, M. J.,M.Mashkoor, Alam. and Irfan, Ahmad. 1990. Nematode Bio-control (Aspects and Prospects)-by CS Publishers & Distributors, New Delhi-110032.

Siddique, S. and Grundler, F. M. 2018. Parasitic nematodes manipulate plant development to establish feeding sites. *Current Opinion of Microbiology.* **46:** 102–108.

Singh, S., Singh, B. and Singh, A. P. 2015. Nematodes: a threat to sustainability of agriculture. *Procedia Environ. Sci.* **29**: 215–216.

R.K.Walia &Harish K.Bajaj. 2003. Textbook of Introductory Plant Nematology-published by ICAR, New Delhi.

Xiang, N., Lawrence, K. S. and Donald, P. A. 2018. Biological control potential of plant growth-promoting rhizobacteria suppression of Meloidogyne incognita on cotton and Heterodera glycines on soybean: *a review. J. Phytopathol.* 166 449–458.

Zhang, S., Gan, Y., Ji, W., Xu, B., Hou, B. and Liu, J. 2017. Mechanisms and characterization of Trichoderma longibrachiatum T6 in suppressing nematodes (Heterodera avenae) in wheat. *Front. Plant Sci.* **8**:1491.

COLOUR PLATES

Coconut black headed caterpillar larvae (p. 69)

Bracon hebetor (p. 69)

Goniozus nephantidis (p. 69)

Coconut black headed caterpillar infested garden (p. 69)

Bracon brevicornis (p. 69)

Coconut slug caterpillar infested leaf (p. 70)

Pediobius imbrues **(p. 70)**

Rhinoceros beetle damage (p. 70)

Red palm weevil damage (p. 70)

Apertochrysa astur **eggs (p. 71)**

Apertochrysa astur **grub (p. 71)**

Fig1: Predator, *Cryptolaemus*,a ladybird beetle feeding on mealybug species. (Courtesy: M. J. Raupp, Cornell University) (p. 71)

Fig 2: Parasitoid *Encarsia* against whitefly. (Courtesy: M. Hoddle, Cornell University) (p. 71)

Stem bleeding disease (p. 72)

Basal stem rot (Ganoderma wilt) disease (p. 72)

Bud rot disease (p. 72)

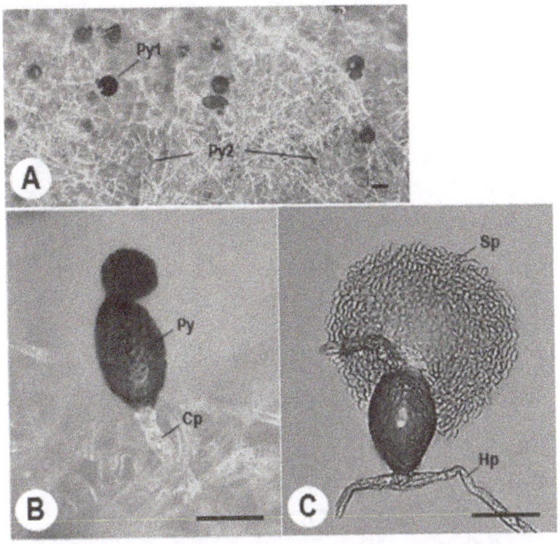

Fig 3: Ampelomyces strain against Powdery mildew pathogen (Courtesy: Nemeth et al., 2021) (p. 73)

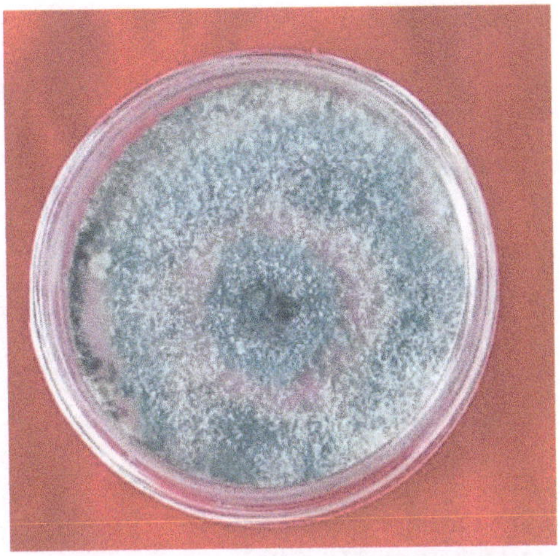

Fig 4: Culture of *Trichodema*, a very promising biological control agent (p. 73)

Indian Rose beetle (*Adoretus versutus*)

(p. 103)

Black leaf chaffer beetle (*Apogonia blanchradii*)

(p. 103)

Bag worm [*Pteroma plagiophelps*]

(p. 103)

Bag worm [*Pteroma plagiophelps*]

(p. 103)

Chaffer beetle damage on cocoa (p. 103)

Bark eating caterpillar (*Indarbela* sp.) (p. 103)

Bark eating caterpillar (*Indarbela* sp.) (p. 103)

***Euproctis fraterna* (p. 103)**

***Euproctis subnotata* (p. 103)**

***Phytophthora* canker (p. 104)**

Pod rot (p. 104)

***Botryodiplodia* pod rot (p. 104)**

Apple and nut borer
(Thylacoptila paurosema)
(p. 113)

Apple and nut borer
(Thylacoptila paurosema)
(p. 113)

Cashew stem and root borer
(Plocaederus ferrugineus)
(p. 113)

Cashew stem and root borer
(Plocaederus ferrugineus)
(p. 113)

Cashew stem and root borer
(Plocaederus ferrugineus)
(p. 113)

Cashew stem and root borer
(Plocaederus ferrugineus)
(p. 113)

Leaf and blossom webber,
Lamida (Macalla) moncusalis
(p. 114)

Leaf and blossom webber,
Lamida (Macalla) moncusalis
(p. 114)

Leaf and blossom webber,
Lamida (Macalla) moncusalis
(p. 114)

Leaf and blossom webber,
Lamida (Macalla) moncusalis
(p. 114)

Leaf miner
Acrocercops syngramma
(p. 114)

Leaf miner
Acrocercops syngramma
(p. 114)

Leaf miner
Acrocercops syngramma
(p. 115)

Shoot tip cater pillar
Anarsia epotias
(p. 115)

Gummosis:
Lasiodiplodia theobraomae
(p. 115)

Gummosis:
Lasiodiplodia theobraomae
(p. 115)

Anthracnose
(Colletotrichum gloeosporiodes)
(p. 115)

Damping off in Cashew seedlings
(p. 115)

Yellowing of guava plant due to nematode infestaion (p. 154)

Root knots (*Meloidogyne* spp) on guava roots (p. 154)

Root galls (*Meloidogyne* spp) on banana roots (p. 154)

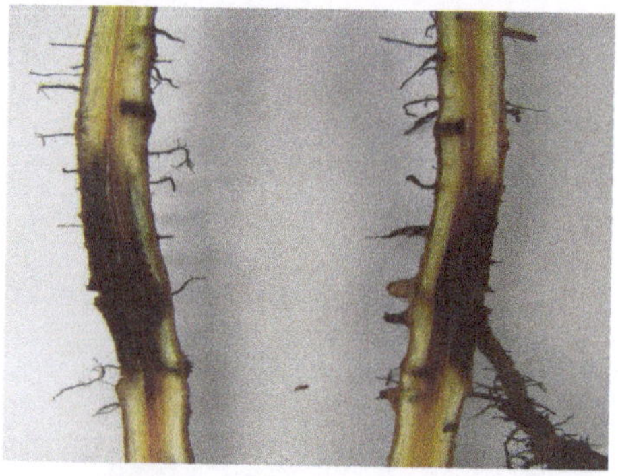

Root lesions (*Pratylenchus* spp) on banana roots (p. 154)

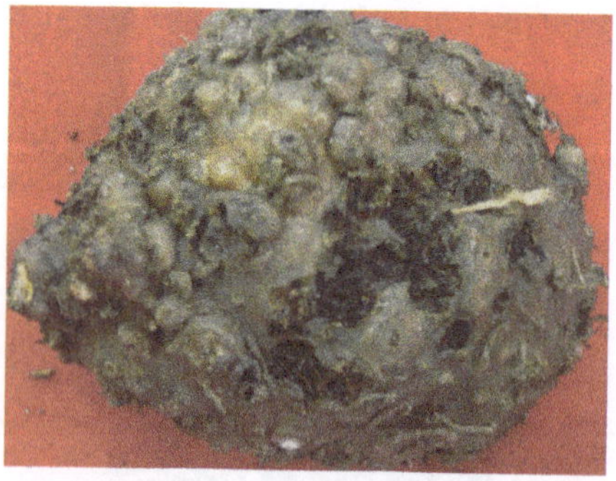

Root knots on Elephant foot yam (p. 154)

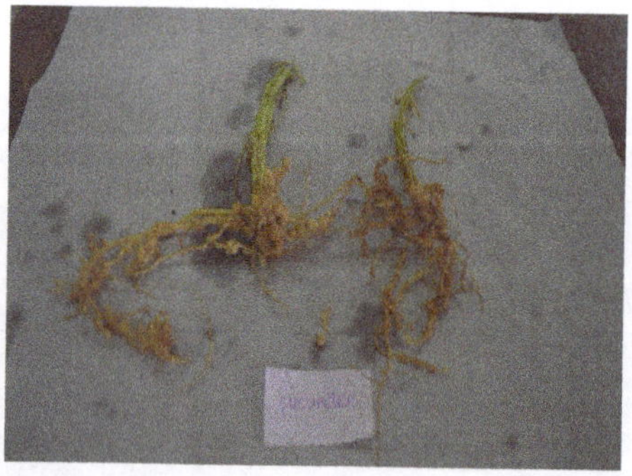

Root galls on cucumber roots (p. 154)

Root knots on tomato (p. 155)

Wilting & drying of pomegranate plant due to nematode incidence (p. 155)

Root galls on crossandra (p. 155)

Root galls on sweet potato (p. 155)

Root galls on Ixora (Ornamental plant) (p. 155)

Index

S

T

U

W

X

Y

Printed at Repro India Ltd.